D1288015

GALILEO GOES TO JAIL

AND OTHER MYTHS ABOUT SCIENCE AND RELIGION

GALILEO GOES TO JAIL

AND OTHER MYTHS ABOUT SCIENCE AND RELIGION

Edited by Ronald L. Numbers

HARVARD UNIVERSITY PRESS
Cambridge, Massachusetts & London, England   2009

Library of Congress Cataloging-in-Publication Data

Galileo goes to jail and other myths about science and religion / edited by Ronald L. Numbers.

   p. cm.

Includes bibliographical references and index.

ISBN 978-0-674-03327-6

   1. Science—History. 2. Scientists—History. 3. Religion and science—History. 4. Religion and state—History. I. Numbers, Ronald L.

Q126.8.G35 2009

215—dc22   2008041250

For

Keith R. Benson

and Carter,

the perfect hosts—

with appreciation from us all

CONTENTS

ACKNOWLEDGMENTS

This book would not exist without the support of a number of individuals and institutions. At the John Templeton Foundation Charles L. Harper, Jr., and Paul K. Wason provided moral and financial backing—and left us free to follow the evidence wherever it led. With the generous assistance of the foundation we were able to bring the contributors together in the summer of 2007 for a working conference at Green College, University of British Columbia, overlooking the Pacific Ocean. Our host there, Keith R. Benson, has been a close collaborator on this project almost from the beginning. During the conference a number of colleagues at the University of British Columbia—John Beatty, Keith Benson, Robert Brain, Alexei Kojevnikov, Adam Shapiro, and Jessica Wang—offered valuable commentary. On occasion we received encouragement and useful input from our distinguished advisory committee: Francisco J. Ayala, John Hedley Brooke, Noah Efron, Ekmeleddin İhsanoğlu, Peter Harrison, David C. Lindberg, Margaret J. Osler, and Nicolaas A. Rupke. Kate Schmit provided indispensable editorial assistance. Ann Downer-Hazell at Harvard University Press was, as usual, the consummate editor. My sincerest thanks to all.

GALILEO GOES TO JAIL

AND OTHER MYTHS ABOUT SCIENCE AND RELIGION

INTRODUCTION

Ronald L. Numbers

I propose, then, to present to you this evening an outline of the great sacred struggle for the liberty of Science—a struggle which has been going on for so many centuries. A tough contest this has been! A war continued longer—with battles fiercer, with sieges more persistent, with strategy more vigorous than in any of the comparatively petty warfares of Alexander, or Cæsar, or Napoleon . . . In all modern history, interference with Science in the supposed interest of religion—no matter how conscientious such interference may have been—has resulted in the direst evils both to Religion and Science, and *invariably.*

> —Andrew Dickson White, "The Battle-Fields of Science" (1869)

The antagonism we thus witness between Religion and Science is the continuation of a struggle that commenced when Christianity began to attain political power . . . The history of Science is not a mere record of isolated discoveries; it is a narrative of the conflict of two contending powers, the expansive force of the human intellect on one side, and the compression arising from traditionary faith and human interests on the other.

> —John William Draper, *History of the Conflict between Religion and Science* (1874)

The greatest myth in the history of science and religion holds that they have been in a state of constant conflict. No one bears more responsibility for promoting this notion than two nineteenth-century American polemicists: Andrew Dickson White

(1832–1918) and John William Draper (1811–1882). White, the young president of Cornell University, became a believer in the warfare between science and religion after religious critics branded him an infidel for, as he put it, trying to create in Ithaca "[a]n asylum for *Science*—where truth shall be sought for truth's sake, not stretched or cut exactly to fit Revealed Religion." On a winter's evening in December 1869 he strode to the podium in the great hall of Cooper Union in New York City, ready to smite his enemies with history, to give them "a lesson which they will remember." In a melodramatic lecture titled "The Battle-Fields of Science" the historian surveyed "some of the hardest-fought battle-fields" of the "great war" between science and religion. He told of Giordano Bruno's being "burned alive as a monster of impiety," of Galileo's having been "tortured and humiliated as the worst of unbelievers," and much more, ending with the latest scientific martyrs, Cornell University and its beleaguered president. As White must have anticipated, his lecture sparked even more controversy, prompting, according to one observer, "instantaneous outcry and opposition." Over the next quarter century White expanded his talk into a huge two-volume work, *A History of the Warfare of Science with Theology in Christendom* (1896), widely translated and frequently reprinted down to the present. In it, as Elizabeth Cady Stanton gleefully noted, he showed "that the Bible has been the greatest block in the way of progress."[1]

Draper was equally exercised when he wrote his *History of the Conflict between Religion and Science* (1874). An accomplished physician, chemist, and historian, Draper largely excused Protestantism and Eastern Orthodoxy of crimes against science while excoriating Roman Catholicism. He did so, he wrote, "partly because its adherents compose the majority of Christendom, partly because its demands are the most pretentious, and partly because it has commonly sought to enforce those demands by the civil power." In addition to chronicling the church's age-old opposition to scientific progress, he ridiculed the recently promulgated doctrine of papal infallibility, which he attributed to men "of sin and shame." He never publicly mentioned, however, what may have agitated him the most: his antipathy toward his own sister, Elizabeth, who had converted to Catholicism and who for a time lived with the Drapers. When one of the Draper children, eight-year-old William, lay near death, Aunt Elizabeth hid his favorite book, a Protestant devotional tract—and did not return it until after the boy had passed away. The grieving father angrily kicked her out of his house, no doubt blaming the Vatican for her un-Christian and dogmatic behavior. Draper's tale of "ferocious theologians" hounding the pioneers of science "with a Bible in one hand and a fiery fagot in the other," as one critic characterized his account, understandably provoked numerous counterattacks. The American convert to Catholicism Orestes Brownson, who described the book as "a tissue of lies from beginning to end," could scarcely contain his fury. "A thousand highway-robberies or a thousand cold-blooded murders," he fumed, "would be but a light social offence in comparison with the publication of one such book as this before us."[2]

Discussions of the relationship between "science" and "religion" originated in the early nineteenth century, when students of nature first began referring to their work as science rather than as natural philosophy (or natural history). Before that time there were occasional expressions of concern about the tension between faith and reason, but no one pitted religion against science or vice versa.[3] By the 1820s, however, books and articles featuring the phrase "science and religion" in their titles were starting to appear. One of the first, if not *the* first, English-language books with the words in their titles came out in 1823: Thomas Dick's popular *The Christian Philosopher; or, The Connection of Science and Philosophy with Religion.* By midcentury "science and religion" was becoming a literary trope, and during the 1850s and 1860s several American colleges and seminaries established professorships devoted to demonstrating (and preserving) the harmony of science and revealed religion.[4]

Although a few freethinkers, most notoriously Thomas Cooper of South Carolina College, denounced religion as "the great enemy of Science," antebellum Americans, especially the clergy, worried far more about the threat of science to orthodox Christianity than about religious barriers to science. By the middle third of the nineteenth century some observers were beginning to suspect that "every new conquest achieved by science, involved the loss of a domain to religion." Especially disturbing were scientific challenges to the first chapters of the Bible. During the three decades between about 1810 and 1840 men of science pushed successfully to replace the supernatural creation of the solar system with the nebular hypothesis, to expand the history of life on earth from 6,000 to millions of years, and to shrink Noah's flood to a regional event in the Near East. Many Christians readily adjusted their reading of the Bible to accommodate such findings, but some biblical literalists thought that the geologists of the day were taking too many liberties with God's word. The Reverend Gardiner Spring, for example, resented scientific efforts to explain creation, which he regarded as "a great *miracle*," incapable of being accounted for scientifically. "The collision is not between the Bible & *Nature*," he declared, "but between the Bible & *natural philosophers*."[5]

At the time it was not uncommon for men of science to engage in biblical exegesis while denying theologians and clergymen the right to monitor science. This practice, along with the increasing marginalization of theologians from the scientific enterprise, galled Charles Hodge, the most eminent Calvinist theologian in midcentury America. Although he continued to venerate men of science who disclosed "the wonderful works of God," by the late 1850s he was growing increasingly frustrated by their tendency to treat theologians who expressed themselves on scientific subjects as "trespassers" who should mind their own business. He attributed the growing "alienation" between men of science and men of the cloth in part to the former's "assumption of superiority" and their practice of stigmatizing their religious critics "as narrow-minded, bigots, old women, Bible worshippers, etc." He resented the lack of respect frequently shown to religious men, who were instructed by their scientific colleagues to quit meddling in science, while they themselves belittled religious beliefs and values. At times Hodge worried that science, devoid of religion, was becoming downright "satanic." He had no doubt that religion was in a "fight for its life against a large class of scientific men."[6]

The spread of "infidel" science—from geology and cosmogonies to biology and anthropology—caused many Christians, both conservatives and liberals, to feel under attack. According to the southern intellectual George Frederick Holmes, "The struggle between science and religion, between philosophy and faith, has been protracted through centuries; but it is only within recent years that the breach has become so open and avowed as to be declared by many to be irreconcilable." Worse yet, even the working classes were joining the fray. As one British writer noted in 1852, "Science is no longer a lifeless abstraction floating above the heads of the multitude. It has descended to earth. It mingles with men. It penetrates our mines. It enters our workshops. It speeds along with the iron courser of the rail."[7]

The debates over Charles Darwin's *On the Origin of Species* (1859), in which the British naturalist sought "to overthrow the dogma of separate creations" and extend the domain of natural law throughout the organic world, signaled a shift in emphasis. Increasingly, scientists, as they were coming to be called, expressed resentment at playing handmaiden to religion. One after another called not only for scientific freedom but also for the subordination of religion—and the rewriting of history with religion as the villain. The most infamous outburst came from the Irish physicist John Tyndall (1820–1893), who in his 1874 Belfast address as president of the British Association for the Advancement of Science thundered:

The impregnable position of science may be described in a few words. We claim, and we shall wrest from theology, the entire domain of cosmological theory. All schemes and systems which thus infringe upon the domain of science must, *in so far as they do this*, submit to its control, and relinquish all thought of controlling it. Acting otherwise proved disastrous in the past, and it is simply fatuous to-day.

Two years later Tyndall wrote a laudatory preface to a British edition of White's *The Warfare of Science*. With such endorsements, the conflict thesis was well on its way toward becoming the historical dogma of the day, at least among intellectuals seeking freedom from religion.[8]

Historians of science have known for years that White's and Draper's accounts are more propaganda than history.[9] (An opposing myth, that Christianity alone gave birth to modern science, is disposed of in Myth 9.) Yet the message has rarely escaped the ivory tower. The secular public, if it thinks about such issues at all, *knows* that organized religion has always opposed scientific progress (witness the attacks on Galileo, Darwin, and Scopes). The religious public *knows* that science has taken the leading role in corroding faith (through naturalism and antibiblicism). As a first step toward correcting these misperceptions we must dispel the hoary myths that continue to pass as historical truths. No scientist, to our knowledge, ever lost his life because of his scientific views, though, as we shall see in Myth 7, the Italian Inquisition did incinerate the sixteenth-century Copernican Giordano Bruno for his heretical *theological* notions.

Unlike the master mythmakers White and Draper, the contributors to this volume have no obvious scientific or theological axes to grind. Nearly half, twelve of twenty-five, self-identify as agnostic or atheist (that is, unbelievers in religion). Among the remaining thirteen there are five mainstream Protestants, two evangelical Protestants, one Roman Catholic, one Jew, one Muslim, one Buddhist—and two whose beliefs fit no conventional category (including one pious Spinozist). Over half of the unbelievers, including me, grew up in devout Christian homes—some as fundamentalists or evangelicals—but subsequently lost their faith. I'm not sure exactly what to make of this fact, but I suspect it tells us something about why we care so much about setting the record straight.

A final word about our use of the word *myth*: Although some of the myths we puncture may have helped to give meaning to the lives of those embracing them, we do not employ the term in its sophisticated academic sense but rather use it as done in everyday conversation—to designate a claim that is false.

THAT THE RISE OF CHRISTIANITY

WAS RESPONSIBLE FOR THE DEMISE

OF ANCIENT SCIENCE

David C. Lindberg

[O]ne finds a combination of factors behind "the closing of the
Western mind": the attack on Greek philosophy by [the apostle]
Paul, the adoption of Platonism by Christian theologians and the
enforcement of orthodoxy by emperors desperate to keep good
order. The imposition of orthodoxy went hand in hand with
a stifling of any form of independent reasoning. By the fifth
century, not only has rational thought been suppressed, but there
has been a substitution for it of "mystery, magic, and authority."

—Charles Freeman, *The Closing of the Western Mind:*
The Rise of Faith and the Fall of Reason (2003)

One spring day in 415, as the story is told, an angry mob of
Christian zealots in Alexandria, Egypt, stirred to action by the recently installed bishop, Cyril, brutally murdered the beautiful,
young pagan philosopher and mathematician Hypatia. Tutored
initially by her father, an accomplished mathematician and astronomer, Hypatia had gone on to write learned commentaries of
her own on mathematical and philosophical texts. Her popularity and influence—and especially her defense of science against
Christianity—so angered the bishop that he ordered her death.
Versions of this story have been a staple of anti-Christian polemics
since the early Enlightenment, when the Irish freethinker John

Toland wrote an overwrought pamphlet, the title of which tells it all: *Hypatia; or, The History of a Most Beautiful, Most Virtuous, Most Learned and in Every Way Accomplished Lady; Who Was Torn to Pieces by the Clergy of Alexandria, to Gratify the Pride, Emulation, and Cruelty of the Archbishop, Commonly but Undeservedly Titled St. Cyril* (1720). According to Edward Gibbon, author of *The History of the Decline and Fall of the Roman Empire* (1776–88), "Hypatia was torn from her chariot, stripped naked, dragged to the church, and inhumanly butchered by the hands of Peter the reader and a troop of savage and merciless fanatics: her flesh was scraped from her bones with sharp oystershells, and her quivering limbs were delivered to the flames." In some accounts Hypatia's murder marked the "death-blow" to ancient science and philosophy. The distinguished historian of ancient science B. L. Van der Waerden claims that "[a]fter Hypatia, Alexandrian mathematics came to an end"; in his study of ancient science, Martin Bernal uses Hypatia's death to mark "the beginning of the Christian Dark Ages."[1]

The story of Hypatia's murder is one of the most gripping in the entire history of science and religion. However, the traditional interpretation of it is pure mythology. As the Czech historian Maria Dzielska documents in a recent biography, Hypatia got caught up in a political struggle between Cyril, an ambitious and ruthless churchman eager to extend his authority, and Hypatia's friend Orestes, the imperial prefect who represented the Roman Empire. In spite of the fact that Orestes was a Christian, Cyril used his friendship with the pagan Hypatia against him and accused her of practicing magic and witchcraft. Although killed largely in the gruesome manner described above—as a mature woman of about sixty years—her death had everything to do with local politics and virtually nothing to do with science. Cyril's crusade against pagans came later. Alexandrian science and mathematics prospered for decades to come.[2]

The misleading accounts of Hypatia's death and Freeman's *Closing of the Western Mind,* quoted above, are attempts to keep alive an old myth: the portrayal of early Christianity as a haven of anti-intellectualism, a fountainhead of antiscientific sentiment, and one of the primary agents responsible for Europe's descent into what are popularly referred to as the "dark ages." Supporting evidence is available, if not plentiful. The apostle Paul (whose influence in shaping Christian attitudes was, of course, enormous) warned the Colossians: "Be on your guard; do not let your minds be captured by hollow and delusive speculations, based on traditions of man-made teaching centered on the elements of the natural world and not on Christ." And in his first letter to the Corinthians, he admonished: "Make no mistake about this: if there is anyone among you who fancies himself wise . . . he must become a fool to gain true wisdom. For the wisdom of this world is folly in God's sight."[3]

Similar sentiments were expressed by several early church fathers, concerned to counter heresy and protect Christian doctrine from the influence of pagan philosophy. The North African Carthaginian Tertullian (ca. 160–ca. 240), a superbly educated and highly influential defender of orthodox Christian doctrine, was undoubtedly the most outspoken of these defenders of Christian orthodoxy. In his most famous utterance, he inquired:

> What indeed has Athens [meant to represent pagan scholarship] to do with Jerusalem [representing Christian religion]? What concord is there between the Academy [presumably Plato's] and the Church? What between heretics and Christians? . . . Away with all attempts to produce a mottled Christianity of Stoic, Platonic, and dialectic composition! We want no curious disputation after possessing Christ Jesus, no inquisition after enjoying the gospel! With our faith, we desire no further belief. For once we believe this, there is nothing else that we ought to believe.[4]

Tertullian's contemporary, Tatian (fl. ca. 172), a Greek-speaking Mesopotamian who made his way to Rome, inquired of the philosophers:

What noble thing have you produced by your pursuit of philosophy? Who of your most eminent men has been free from vain boasting? . . . I could laugh at those who in the present day adhere to [Aristotle's] tenets—people who say that sublunary things are not under the care of Providence . . . Wherefore be not led away by the solemn assemblies of philosophers who are no philosophers, who dogmatize the crude fancies of the moment.[5]

Similar complaints were voiced by other critics of pagan (that is, non-Christian) learning.

But to stop here would be to present a seriously incomplete and highly misleading picture. The very writers who denounced Greek philosophy also employed its methodology and incorporated large portions of its content in their own systems of thought. From Justin Martyr (d. ca. 165) to Saint Augustine (354–430) and beyond, Christian scholars allied themselves with Greek philosophical traditions deemed congenial to Christian thought. Chief among these philosophies was Platonism (or Neoplatonism), but borrowing from Stoic, Aristotelian, and neo-Pythagorean philosophy was also common. Even the denunciations issuing from Christian pens, whether of specific philosophical positions or of philosophy in general, often reflected an impressive command of Greek and Roman philosophical traditions.

But what did these religious and philosophical traditions have to do with *science?* Was there any activity or body of knowledge at the time that can be identified as "science"? If not, then the myth, as stated, is obviously false. But let us not allow ourselves to escape so easily. In the period that we are discussing, there *were* inherited beliefs about nature—about the origins and structure of the cosmos, the motions of celestial bodies, the nature of the elements, sickness and health, the explanation of dramatic natural phenomena (thunder, lightning, eclipses, the rainbow, and the like)—and its relationship to the gods. These are the ingredients of what would develop centuries later into modern science (some were already identical to their modern counterparts); and if we are interested in the origins of Western science they are what we must investigate. For the naming of these enterprises, historians of science have chosen a variety of expressions—"natural philosophy" and "mathematical science" being the most common. For the sake of clarity, I choose to refer to them simply as the "classical sciences"—that is, the sciences that descended from the Greek and Roman classical tradition—and to their practitioners as "scientists" or "philosopher/scientists."

As we have seen, Christian writers sometimes expressed deep hostility toward the classical sciences. Tertullian, whom we have already met, attacked pagan philosophers for their assignment of divinity to the elements and the sun, moon, planets, and stars. In the course of his argument, he vented his wrath over the vanity of the ancient Greek scientist/philosophers:

> Now pray tell me, what wisdom is there in this hankering after conjectural speculations? What proof is afforded to us . . . by the useless affectation of a scrupulous curiosity, which is tricked out with an artful show of language? It therefore served Thales of Miletus [philosopher of the 6th c. B.C.] quite right, when, star-gazing as he walked . . . , he had the mortification of falling into a well . . . His fall, therefore, is a figurative picture of the philosophers; of those, I mean, who persist in applying their studies to a vain purpose, since they indulge a stupid curiosity on natural objects.[6]

But it was an *argument* that Tertullian presented, and to a very significant degree he built it out of materials and by the use of methods drawn from the Greco-Roman philosophical tradition. He argued, for example, that the precise regularity of the orbital motions of the celestial bodies (a clear reference to the findings of Greek astronomers) bespeaks a "governing power" that rules over them; and if they are ruled over, they surely cannot be gods. He also introduced the "enlightened view of Plato" in support of the claim that the universe must have had a beginning and therefore cannot itself partake of divinity; and in this and other works he "triumphantly parades" his learning (as one of his biographers puts it) by naming a long list of other ancient authorities.[7]

Basil of Caesarea (ca. 330–379), representing a different century and a different region of the Christian world, revealed similar attitudes toward the classical sciences. He sharply attacked philosophers and astronomers who "have wilfully and voluntarily blinded themselves to knowledge of the truth." These men, he continued, have "discovered everything, except one thing: they have not discovered the fact that God is the creator of the universe."[8] Elsewhere he inquired why we should "torment ourselves by refuting the errors, or rather the lies of the Greek philosophers, when it is sufficient to produce and compare their mutually contradictory books."[9]

But while attacking the errors of Greek science and philosophy—and what he did not find erroneous, he generally judged useless—Basil also revealed a solid mastery of their contents. He argued against Aristotle's fifth element, the quintessence; he recounted the Stoic theory of cyclic cosmological conflagration and regeneration; he applauded those who employ the laws of geometry to refute the possibility of multiple worlds (a clear endorsement of Aristotle's argument for the uniqueness of the cosmos); he derided the Pythagorean notion of music of the planetary spheres; and he proclaimed the vanity of mathematical astronomy.

Tertullian, Tatian, and Basil have thus far been portrayed as outsiders to the classical tradition, attempting to discredit and destroy what they regarded as a menace to orthodox Christianity. Certainly some of their rhetoric supports such an interpretation, as when they appealed for simple faith as an alternative to philosophical reasoning. But we need to look beyond rhetoric to actual practice; it is one thing to deride the classical sciences and the philosophical systems that undergirded them, or declare them useless, another to abandon them. Despite their derision, Tertullian, Basil, and others like them were continuously engaged in serious philosophical argumentation, borrowing from the very tradition that they despised. It is no distortion of the evidence to see them as insiders to this tradition, attempting to formulate an alternative philosophy based on Christian principles—opposed not to the enterprise of philosophy but to specific philosophical principles that they considered erroneous and dangerous.

The most influential of the church fathers and the one who most powerfully shaped the codification of Christian attitudes toward nature was Augustine of Hippo (354–430). Like his predecessors, Augustine had serious reservations about the value of classical philosophy and science and the legitimacy of their pursuit. But his criticism was muted and qualified by an acknowledgment, in both word and deed, of legitimate uses to which knowledge of the cosmos might be put, including religious utility. In short, although Augustine did not devote himself to promotion of the sciences, neither did he fear them in their pagan versions to the degree that many of his predecessors had.

Scattered throughout Augustine's voluminous writings are worries about pagan philosophy and its scientific partner, and admonitions to Christians not to overvalue them. In his *Enchiridion,* he assured his reader that there is no need to be

> dismayed if Christians are ignorant about the properties and the number of the basic elements of nature, or about the motion, order, and deviations of the stars, the map of the heavens, the kinds and nature of animals, plants, stones, springs, rivers, and mountains . . . For the Christian, it is enough to believe that the cause of all created things . . . is . . . the goodness of the Creator.[10]

In *On Christian Doctrine,* Augustine commented on the uselessness and vanity of astronomical knowledge:

> Although the course of the moon . . . is known to many, there are only a few who know well the rising or setting or other movements of the remainder of the stars without error. Knowledge of this kind in itself, although it is not allied with any superstition, is of very little use in the treatment of the Divine Scriptures and even impedes it through fruitless study; and since it is associated with the most per-

nicious error of vain [astrological] prediction it is more appropriate and virtuous to condemn it.[11]

And finally, in his *Confessions* he argued that "because of this disease of curiosity . . . men proceed to investigate the phenomena of nature, . . . though this knowledge is of no value to them: for they wish to know simply for the sake of knowing."[12] Knowledge for the sake of knowing is without value and, therefore, to be repudiated.

But once again this is not the whole story. Christian philosophers of the patristic period may not have valued philosophy or the sciences for their *intrinsic* value, but from this we cannot conclude that they denied the sciences all *extrinsic* value. For Augustine, knowledge of natural phenomena acquired value and legitimacy insofar as it served other, higher purposes. The most important such purpose is biblical exegesis, since ignorance of mathematics and natural history (zoology and botany) renders us incapable of grasping the literal sense of Scripture. For example, only if we are familiar with serpents will we grasp the meaning of the biblical admonition to "be as wise as serpents and as innocent as doves" (Matthew 10:16). Augustine also conceded that portions of pagan knowledge, such as history, dialectic, mathematics, the mechanical arts, and "teachings that concern the corporeal senses," contribute to the necessities of life.[13]

In his *Literal Commentary on Genesis,* where he put his own superb grasp of Greek cosmology and natural philosophy to good use, Augustine expressed dismay at the ignorance of some Christians:

Even a non-Christian knows something about the earth, the heavens, and the other elements of this world, about the motion and orbit of the stars and even their size and relative positions, about the predictable eclipses of the sun and moon, the cycles of the years and the seasons, about the kinds of animals, shrubs, stones, and so forth, and this knowledge he holds to, as being certain from reason and experience. Now it is a disgraceful and dangerous thing for an infidel [a non-Christian] to hear a Christian . . . talking nonsense on these topics; and we should take all means to prevent such an embarrassing situation, in which people show up vast ignorance in a Christian and laugh it to scorn.[14]

Insofar as we require philosophical or scientific knowledge of natural phenomena—and Augustine is certain that we do—we must take them from the people who possess it: "If those who are called philosophers, especially the Platonists, have said things which are indeed true and are well accommodated to our faith, they should not be feared; rather, what they have said should be taken from them as from unjust possessors and converted to our use."[15] All truth is ultimately God's truth, even if found in the books of pagan authors; and we should seize it and use it without hesitation.

In Augustine's influential view, then, knowledge of the things of this world is not a legitimate end in itself, but as a means to other ends it is indispensable. The classical sciences must accept a subordinate position as the handmaiden of theology and religion— the temporal serving the eternal. The knowledge contained in classical sciences is not to be loved, but it may legitimately be used. This attitude toward scientific knowledge came to prevail throughout the Middle Ages and survived well into the modern period. Augustine's handmaiden science was defended explicitly and at great length, for example, by Roger Bacon in the thirteenth century, whose defense of useful knowledge contributed to his notoriety as one of the founders of experimental science.[16]

Does endowing scientific knowledge with handmaiden status constitute a serious blow against scientific progress? Are the critics of the early church right in viewing it as the opponent of genuine science? I would like to make three points in reply. (1) It is certainly true that the fathers of the early Christian church did not view support of the classical sciences as a major obligation. These sciences had low priority for the church fathers, for whom the major concerns were (quite properly) the establishment of Christian doctrine, defense of the faith, and the edification of believers. But (2), low or medium priority was far from zero priority. Throughout the Middle Ages and well into the modern period the handmaiden formula was employed countless times to justify the investigation of nature. Indeed, some of the most celebrated achievements of the Western scientific tradition were made by religious scholars who justified their labors (at least in part) by appeal to the handmaiden formula. (3) No institution or cultural force of the patristic period offered more encouragement for the investigation of nature than did the Christian church. Contemporary pagan culture was no more favorable to disinterested speculation about the cosmos than was Christian culture. It follows that the presence of the Christian church enhanced, rather than damaged, the development of the natural sciences.

But we must not forget Tertullian and his fiery opposition to the classical sciences. Did he not represent a substantial group of outspoken opponents of the classical sciences? Not as far as the historical record reveals. One must work hard to find suitable passages from the writings of Tatian, Basil, and others in denigration of the classical philosophy. And even then their rhetoric was many decibels below that of Tertullian; moreover, their opposition was to aspects of classical tradition that had little to do with the classical sciences. Scores of church fathers and their counterparts in later centuries wrestled with aspects of classical philosophy, attempting to reconcile it with biblical teachings and orthodox Christian theology; but when it came to the classical sciences, the great majority joined Augustine: approach the classical sciences with caution; fear them if you must, but put them to work as the handmaidens of Christian philosophy and theology if you can. So, to put it bluntly, the scholars wishing to demonstrate Christian hostility toward the classical sciences built their case on Tertullian because he was their only relevant, sufficiently hostile, exhibit. It was Augustine's sympathetic voice that prevailed in the practice of the sciences from the patristic period, through the Middle Ages, and beyond.

Did Augustine practice what he preached? That he did is best illustrated in his *Literal Commentary on Genesis*, where he produced a verse-by-verse interpretation of the biblical account of creation as it appears in the first three chapters of Genesis. In the course of this work of his mature years, Augustine made copious use of the natural sciences contained in the classical tradition to explicate the creation story. Here we encounter Greco-Roman ideas about lightning, thunder, clouds, wind, rain, dew, snow, frost, storms, tides, plants and animals, matter and form, the four elements, the doctrine of natural place, seasons, time, the calendar, the planets, planetary motion, the phases of the moon, astrological influence, the soul, sensation, sound, light and shade, and number theory. For all of his worry about overvaluing the Greek scientific/philosophical tradition, Augustine and others like him applied Greco-Roman natural science with a vengeance to biblical interpretation. The sciences are not to be loved, but to be used. This attitude toward scientific knowledge was to flourish throughout the Middle Ages and well into the modern period. Were it not for this outlook, medieval Europeans would surely have had less scientific knowledge, not more.

THAT THE MEDIEVAL CHRISTIAN CHURCH
SUPPRESSED THE GROWTH OF SCIENCE

Michael H. Shank

> The Christian party [in the early Middle Ages] asserted that all
> knowledge is to be found in the Scriptures and in the traditions
> of the Church . . . The Church thus set herself forth as the
> depository and arbiter of knowledge; she was ever ready to resort
> to the civil power to compel obedience to her decisions. She thus
> took a course which determined her whole future career: she
> became a stumbling-block in the intellectual advancement of
> Europe for more than a thousand years.
>
> —John William Draper, *History of the Conflict between
> Religion and Science* (1874)

The myth of the medieval church's opposition to science is not
likely to go away—in part because it dovetails so nicely with other
cherished myths about the Middle Ages, in part because it is
so easy to manufacture. Anyone who has heard of Tertullian's
challenge—"What has Athens to do with Jerusalem?"—and of
Galileo's appearance before the Inquisition may simply join these
two points with a straight line. All one needs is the assumption,
also mythical, that Galileo was condemned by a medieval church
doing what it did best. (In fact, as explained in Myth 8, it was the
early-modern Catholic church that censured Galileo, using a new
literalist view of Scripture that would have surprised Augustine
and Thomas Aquinas.)

The crude concept of the Middle Ages as a millennium of stagnation brought on by Christianity has largely disappeared among scholars familiar with the period, but it remains vigorous among popularizers of the history of science—perhaps because, instead of consulting scholarship on the subject, the more recent popularizers have relied upon their predecessors uncritically.

Consider the following claim from a book by Robert Wilson recently published by Princeton University Press. It quotes from Tertullian (ca. 160–ca. 220)

> to illustrate the point that the Christian religion developed on the basis that the Gospel was the primary source of guidance and of truth and was inviolate. This commitment to Holy Scripture was, and still is, the fundamental basis of Christianity, but there is no doubt that it was a discouragement to scientific endeavours and these languished for a thousand years after the military fall of Rome. During that time, possibly because the Gospel was based on ancient writings, other ancient works of a non-religious character, including the writings on science by the ancient Greeks, also came to be regarded as inviolate. These factors were to lead to one of the most unfortunate events in the history of Christianity and science—the trial of Galileo.[1]

Wilson's book has no footnotes: did he perhaps consult the astronomer Carl Sagan's *Cosmos* (1980), a popular predecessor of Wilson's book? This companion to the *Cosmos* film series aired by PBS ends with a timeline of individuals with astronomical associations. It is famous among medievalists for covering Greek antiquity (from Thales to Hypatia), then leaving a thousand years blank and starting again with Leonardo and Copernicus. The caption refers to the empty space as "a poignant lost opportunity for mankind."[2] The power of the myth is such that Sagan does not need to say where the blame lies. Sagan, in turn, may have taken a cue from Henry Smith Williams's *Great Astronomers* (Simon and Schuster, 1930), whose medieval chapter consists of two biblical epigraphs ascribed to an "oriental anthology" followed by several blank pages. This passive form of the myth simply assumes that the medieval answer to Tertullian's question was that Athens had nothing to do with Jerusalem (see Myth 1). Since only Jerusalem mattered, no one bothered with Athens (or Alexandria).

In the more active form of the myth, the medieval church takes specific steps to curtail scientific inquiries: it jails Roger Bacon (ca. 1214–1294), portrayed as the most creative scientist of the era, for two, ten, fourteen, or fifteen years, depending on your web source. The assertion that Bacon was imprisoned (allegedly by the head of his own Franciscan order) first originates some eighty years after his death and has drawn skepticism on these grounds alone. Scholars who find this assertion plausible connect it with Bacon's attraction to contemporary prophecies that have nothing to do with Bacon's scientific, mathematical, or philosophical writings.[3]

Historians of science have presented much evidence against the myth, however. John Heilbron, no apologist for the Vatican, got it right when he opened his book *The Sun in the Church* with the following words: "The Roman Catholic Church gave more financial and social support to the study of astronomy for over six centuries, from the recovery of ancient learning during the late Middle Ages into the Enlightenment, than any other, and probably all, other institutions."[4] Heilbron's point can be generalized far beyond astronomy. Put succinctly, the medieval period gave birth to the university, which developed with the active support of the papacy. This unusual institution sprang up rather spontaneously around famous masters in towns like Bologna, Paris, and Oxford before 1200. By 1500, about sixty universities were scattered throughout Europe. What is the significance of this development for our myth? About 30 percent of the medieval university curriculum covered subjects and texts concerned with the natural world.[5] This was not a trivial development. The proliferation of universities between 1200 and 1500 meant that hundreds of thousands of students—a quarter million in the German universities alone from 1350 on—were exposed to science in the Greco-Arabic tradition. As the universities matured, the curriculum came to include more works by Latin masters who developed this tradition along original lines.

If the medieval church had intended to discourage or suppress science, it certainly made a colossal mistake in tolerating—to say nothing of supporting—the university. In this new institution, Greco-Arabic science and medicine for the first time found a permanent home, one that—with various ups and downs—science has retained to this day. Dozens of universities introduced large numbers of students to Euclidean geometry, optics, the problems of generation and reproduction, the rudiments of astronomy, and arguments for the sphericity of the earth. Even students who did not complete their degrees gained an elementary familiarity with natural philosophy and the mathematical sciences and imbibed the naturalism of these disciplines. This was a cultural phenomenon of the first order, for it affected a literate elite of several hundred thousand students: in the middle of the fifteenth century, enrollments in universities in Germanic territories that have survived to this day (places like Vienna, Heidelberg, and Cologne) reached levels unmatched until the late nineteenth and early twentieth centuries.[6]

But, some would argue, weren't most students monks or priests who spent most of their time studying theology, the queen of the sciences? If all scholars were theologians, doesn't that pretty much say it all? This is another collection of myths. Most students never got close to meeting the requirements for studying theology (usually a master of arts degree). They remained in the faculties of arts, where they studied only nonreligious subjects, including logic, natural philosophy, and the mathematical sciences. In fact, as a result of quarrels between faculties, students in the arts faculty were not allowed to treat theological subjects. In short, most students had no theological or biblical studies at all.

Moreover, not all universities had a faculty of theology. Very few had one in the thirteenth century, and the newer foundations initially were not allowed to have one. By the later Middle

Ages, the papacy permitted more faculties of theology. During the Great Schism, when two popes who had excommunicated each other were competing for the allegiance of the various political rulers, they granted faculties of theology to some universities, like Vienna, that had not had one before. Even so, only a small minority of students ever studied theology, which was the smallest of the three higher faculties in the northern universities. By far the most popular advanced subject was law, which promised careers in the growing bureaucracies of both the church and the secular rulers.

As to theology being the queen of the sciences, this notion goes back to Aristotle—no Christian theologian—who meant by it that metaphysics or theology (as the "science of being") was a branch of philosophy more fundamental than either mathematics or natural philosophy (his two other theoretical "sciences"). While many medieval scholars conceded the great dignity of theology, its scientific status was contested, not least by theologians. Robert Grosseteste (d. 1253), chancellor of Oxford and bishop of Lincoln, held that, for an intellect unburdened by a physical body, theology offered a higher degree of certainty than did mathematics and natural philosophy, but for us mortals here below, mathematics yielded greater certainty.[7] Using Aristotle's criteria, the great Italian theologian and philosopher Thomas Aquinas (ca. 1225–1274) later argued that theology was a science.[8] But not everyone agreed with Aquinas. William of Ockham (ca. 1287–1347), an influential English Franciscan, *denied,* also on Aristotelian grounds, that theology was a science. He noted that the principles of a science must be better known than its conclusions. But the principles of theology are the articles of faith, which, as Ockham was fond of pointing out, often appear "false to all, or to the majority, or to the wisest."[9] Theology therefore did not qualify as a science.

Finally, most students and masters were neither priests nor monks, which required special vows. They did have clerical status, however, at least in northern universities like Paris. This was a hard-won legal category that carried almost no formal obligations, religious or otherwise (students could marry, for example), while conferring one important privilege: the right, resented by the city folk, to be tried in a more lenient university or ecclesiastical court instead of a secular one. This status came in very handy when a student killed a townsman in a barroom brawl. (At Paris, students won this right after going on strike following just such an incident.) Although they were not the majority of the students, many of the best-known writers on natural philosophy and practitioners of the mathematical sciences of the era were churchmen or friars.

Does the myth get a new lease on life if I reveal that lectures on Aristotle's natural philosophy were forbidden at Paris in 1210 (under penalty of excommunication) and in 1215 (under no specified penalty)? It does not. While churchmen acting in their official capacities did issue these condemnations, it is misleading to say that "the Church" did so, for this seems to imply that they were valid for all of Christendom. In each case, however, the condemnations were local, issued by the bishops in a province or by a cardinal legate in relation to Paris.[10] Medieval hairsplitting, you say? Not at all: the point of this qualification is absolutely crucial. To make "the Church" the agent in cases where the condemnation is local is technically correct but highly misleading, for such injunctions affected only a minuscule fraction of the population, and usually not for long. These condemnations did not pertain to students and masters elsewhere. Early-thirteenth-century Oxford, for example, saw no prohibitions of this sort (indeed, the reception of Aristotle at Oxford was very smooth).

It is not clear that the condemnations mattered much, or for long, to people in the affected diocese (mostly that of Paris). Despite the condemnation of 1215, we know that Roger Bacon was teaching Aristotle's *Physics* at Paris in the 1240s. What is more, by 1255 Aristotle's formerly condemned natural-philosophical treatises were *required* for the bachelor's and master's degrees in arts at Paris, as they were already or would be for most medieval universities. Keep in mind, though, that Paris was not typical: it faced many more episcopal condemnations than the average university, and for perfectly good local reasons. Most universities were not subject to this kind of interference.

What was the impact of such condemnations on the pursuit of science in medieval Europe? It was minimal, for one very simple reason: If condemnations were usually tied to one locality, students and masters were not. They could pack up and go elsewhere, and they did. Indeed, when in 1229–1231 the university of Paris went on strike on account of a conflict with local authorities, the university of Toulouse invited the Parisian students to travel south ("the second land of promise, flowing with milk and honey . . . Bacchus reigns in vineyards") and reminded them that Toulouse had no ban on Aristotle ("Those who wish to scrutinize the bosom of nature to the innermost can hear here the books of Aristotle that were forbidden at Paris").[11] Paris, the "new Athens," soon reopened thanks to the papal bull *Parens scientiarum* ("mother of the sciences"), which mostly upheld the masters' privileges against the bishop.[12]

Ah, you say, but what about 1277, when "the Church" condemned 219 academic propositions, again in Paris? This most famous of medieval condemnations attacked astrological determinism, a number of Aristotelian theses (including the impossibility of a vacuum), and such humorous or self-serving theses as "The only wise men in the world are philosophers" and "Nothing is known better by knowing theology."[13] Again, this condemnation was issued by the bishop of Paris, aided by some conservative theologians at the university; they used the occasion to clamp down on uppity philosophers and to lash out at their Aristotelian colleague, Thomas Aquinas. Ironically, a century ago the historian Pierre Duhem credited this condemnation with a very positive effect on science. He argued that it forced philosophers to free themselves from their fondness for Aristotle's theses and to consider alternatives. For him, the date 1277 thus marked the beginning of modern (i.e., non- or anti-Aristotelian) science. Nowadays, however, historians agree that this is too great a burden for the Parisian condemnations of 1277 to bear.

A short list of accomplishments from the period suggests that the inquiry into nature did not stagnate in medieval Europe. In the late thirteenth century, William of Saint-Cloud pioneered the use of the camera obscura to view solar eclipses. In the early fourteenth century, Dietrich von Freiberg (a Dominican) solved the problem of the primary and secondary rainbows: he appealed, respectively, to one and two internal reflections inside the raindrop, which he modeled using a glass vial filled with water. Meanwhile, at Oxford, natural philosophers were applying mathematical analysis to motion, coming up with theoretical ways of measuring uniformly changing quantities. In mid-fourteenth-century Paris, Jean Buridan used impetus theory to explain projectile motion, the acceleration of free-fall, and even the unceasing rotation of the starry sphere (in the absence of resistance, God's initial impetus at creation is preserved and requires no further intervention). His younger contemporary Nicole Oresme (later a bishop) offered a nice list of arguments for the possible rotation of the earth: he concluded that no available empirical or rational evidence could determine whether or not it moved. Many more examples could be cited. Like most masters, these individuals benefited from the considerable freedom of thought allowed by the university disputation, which required that arguments pro and contra various positions be advanced and defended on rational grounds alone. It was the scholars' fellow disputants who regularly sought to give them grief; most of the time, "the Church" did not.

Between 1150 and 1500, more literate Europeans had had access to scientific materials than any of their predecessors in earlier cultures, thanks largely to the emergence, rapid growth, and naturalistic arts curricula of the medieval universities. If the medieval church had intended to suppress the inquiry into nature, it must have been completely powerless, for it utterly failed to reach its goal.

THAT MEDIEVAL CHRISTIANS TAUGHT
THAT THE EARTH WAS FLAT

Lesley B. Cormack

In Christendom, the greater part of this long period [Ptolemy to Copernicus] was consumed in disputes respecting the nature of God, and in struggles for ecclesiastical power. The authority of the Fathers, and the prevailing belief that the Scriptures contain the sum of all knowledge, discouraged any investigation of Nature . . . This indifference continued until the close of the fifteenth century. Even then there was no scientific inducement. The inciting motives were altogether of a different kind. They originated in commercial rivalries, and the question of the shape of the earth was finally settled by three sailors, Columbus, Da Gama, and above all, by Ferdinand Magellan.

—John William Draper, *History of the Conflict between Religion and Science* (1874)

With the decline of Rome and the advent of the Dark Ages, geography as a science went into hibernation, from which the early Church did little to rouse it . . . Strict Biblical interpretations plus unbending patristic bigotry resulted in the theory of a flat earth with Jerusalem in its center, and the Garden of Eden somewhere up country, from which flowed the four Rivers of Paradise.

—Boise Penrose, *Travel and Discovery in the Renaissance* (1955)

A Europe-wide phenomenon of scholarly amnesia . . . afflicted the continent from AD 300 to at least 1300. During those centuries Christian faith and dogma suppressed the useful image

of the world that had been so slowly, so painfully, and so
scrupulously drawn by ancient geographers.
—Daniel J. Boorstin, *The Discoverers* (1983)

Did people in the Middle Ages think that the world was flat?
Certainly the writers quoted above would make us think so. As
the story goes, people living in the "Dark Ages" were so ignorant (or so deceived by Catholic priests) that they believed the
earth was flat. For a thousand years they lingered in ignorant ob-
scurity, and were it not for the heroic bravery of Christopher
Columbus and other explorers, they might well have continued
in this ignorance for even longer. Thus, it was the innovation
and courage of investors and explorers, motivated by economic
goals and modern curiosity, that finally allowed us to break free
from the shackles forged by the medieval Catholic church.[1]
Where does this story come from? In the nineteenth century,
scholars interested in promoting a new scientific and rational
view of the world claimed that ancient Greeks and Romans had
understood that the world was round, but that this knowledge
was suppressed by medieval churchmen. Pro-Catholic scholars
responded by making the argument that medieval thinkers did
know the world was round.[2] Critics, however, dismissed such
opinions as mere apologetics. Why did the battle rage over this
particular issue? Because a belief in the flat earth was equated
with willful ignorance, while an understanding of the spherical
earth was seen as a measure of modernity; the side one defended
became a means of condemning or praising medieval churchmen. For scholars such as William Whewell or John Draper,
therefore, Catholicism was bad (since it promoted a flat-earth
view), while for Roman Catholics, Catholicism was good (since
it promoted modernity). As we'll see, neither of these extremes
describes the true state of affairs.[3]
This equation of rotundity with modernity also explains why
nineteenth-century American historians claimed it was Columbus

and the early mercantilists who proved the earth was round and thereby ushered in modernity—and America. In fact, it was a biography of Columbus by the American author Washington Irving, the creator of "Rip Van Winkle," that introduced this idea to the world.[4]

But the reality is more complex than either of these stories. Very few people throughout the Middle Ages believed that the world was flat. Thinkers on both sides of the question were Catholics, and for them, the shape of the earth did not equate with progressive or traditionalist views. It is true that most clerics were more concerned with salvation than the shape of the earth—that was their job, after all. But God's works in nature were important to them as well. Columbus could not have proved that the world was round, because this fact was already known. Nor was he a rebellious modern—he was a good Catholic and undertook his voyage believing he was doing God's work. A transformation was taking place in fifteenth-century views of the earth, but it had more to do with a new way of mapping than with a move from flat earth to round sphere.

Scholars in antiquity developed a very clear spherical model of the earth and the heavens. Every major Greek geographical thinker, including Aristotle (384–322 B.C.), Eratosthenes (third century B.C.), and Ptolemy (second century A.D.), based his geographical and astronomical work on the theory that the earth was a sphere. Likewise, all of the major Roman commentators, including Pliny the Elder (23–79 A.D.), Pomponius Mela (first century A.D.), and Macrobius (fourth century A.D.), agreed that the earth must be round. Their conclusions were in part philosophical—a spherical universe required a sphere in the middle—but were also based on mathematical and astronomical reasoning.[5] Most famous was Aristotle's proof of the sphericity of the earth, an argument used by many thinkers in the Middle Ages and Renaissance.

If we examine the work of even early-medieval writers, we find that with few exceptions they held a spherical-earth theory.

Among the early church fathers, Augustine (354–430), Jerome (d. 420), and Ambrose (d. 420) all agreed that the earth was a sphere. Only Lactantius (early fourth century) provided a dissenting opinion, but he rejected all pagan learning since it distracted people from their real work of achieving salvation.[6]

From the seventh century to the fourteenth, every important medieval thinker concerned about the natural world stated more or less explicitly that the world was a round globe, many of them incorporating Ptolemy's astronomy and Aristotle's physics into their work. Thomas Aquinas (d. 1274), for example, followed Aristotle's proof in demonstrating that the changing positions of the constellations as one moved about on the earth's surface indicated the spherical shape of the earth. Roger Bacon (d. 1294), in his *Opus Maius* (ca. 1270), stated that the world was round, that the southern antipodes were inhabited, and that the sun's passage along the line of the ecliptic affected climates of different parts of the world. Albertus Magnus (d. 1280) agreed with Bacon's findings, while Michael Scot (d. 1234) "compared the earth, surrounded by water, to the yolk of an egg and the spheres of the universe to the layers of an onion."[7] Perhaps the most influential were Jean de Sacrobosco, whose *De Sphera* (ca. 1230) demonstrated that the earth was a globe, and Pierre d'Ailly (1350–1410), archbishop of Cambrai, whose *Imago Mundi* (written in 1410) discussed the sphericity of the earth.[8] Both of these books enjoyed great popularity; Sacrobosco's book was used as a basic textbook throughout the Middle Ages, while d'Ailly's book was read by early explorers like Columbus.

The one medieval author whose work has sometimes been interpreted to demonstrate belief in a disk-shaped rather than spherical earth is Isidore of Seville (570–636), a prolific encyclopedist and natural philosopher. Although he was explicit about the spherical shape of the universe, historians have remained divided on his portrayal of the shape of the earth itself.[9] He claimed that everyone experienced the size and heat of the sun in the same manner, which could be interpreted to mean that sunrise was seen at the same moment by all the earth's inhabitants and that therefore the earth was flat; but the statement more likely implies that the sun's shape did not alter as it progressed around the earth. Much of his physics and astronomy can only be understood to depend on a spherical earth, as does his interpretation of lunar eclipses. While it is not necessary to insist on absolute consistency, it does seem that Isidore's cosmology is only consistent with a spherical earth.[10]

Many popular vernacular writers in the Middle Ages also supported the idea of a round earth. Jean de Mandeville's *Travels* to the Holy Land and to the Earthly Paradise beyond, written in about 1370, was one of the most widely read books in Europe from the fourteenth to the sixteenth century. Mandeville was quite explicit in stating that the world was round and navigable:

> And therefore I say sickerly that a man myght go all the world about, both above and beneath, and come again to his own country . . . And alway he should find men, lands, isles and cities and towns, as are in their countries.[11]

Likewise, Dante (1265–1321) in the *Divine Comedy* described the world as a sphere several times, claiming that the southern hemisphere was covered with a vast sea. And in "The Franklin's Tale" Chaucer (ca. 1340–1400) spoke of "This wyde world, which that men seye is round."[12]

The one medieval writer explicitly to deny the sphericity of the earth was Cosmas Indicopleustes, a sixth-century Byzantine monk who may have been influenced by contemporary Jewish and Eastern flat-earth traditions. Cosmas developed a scripturally based cosmology, with the earth as a tableland, or plateau, placed at the bottom of the universe. It is hard to know how influential he was during his lifetime. Only two copies of his treatise exist today, one of which may have been Cosmas's personal copy, and only one man in the Middle Ages is known to have read his work, Photius of Constantinople (d. 891), widely regarded as the best-read man of his age.[13] In the absence of positive evidence, we cannot use

Cosmas to argue that the Christian church suppressed knowledge of the rotundity of the earth. Cosmas's work merely indicates that the early-medieval scholarly climate was open to debates on the subject.

With the exceptions of Lactantius and Cosmas, all major scholars and many vernacular writers interested in the physical shape of the earth, from the fall of Rome to the time of Columbus, articulated the theory that the earth was round. The scholars may have been more concerned with salvation than with geography, and the vernacular writers may have displayed little interest in philosophical questions. But, with the exception of Cosmas, no medieval writer denied that the earth was spherical—and the Catholic church never took a stand on the issue.

Given this background, it would be silly to argue that Columbus proved the world was round—or even argued so. However, popular accounts continue to circulate the erroneous story that Columbus fought the prejudiced and ignorant scholars and clerics at Salamanca, the home of Spain's leading university, before convincing Queen Isabella to let him try to prove his position. Columbus's proposal—that the distance from Spain west to China was not prohibitively great and that it was shorter and safer than going around Africa—was greeted with incredulity by the group of scholars informally assembled to advise the king and queen of Spain. Since no records remain of that meeting, we must rely on reports written by Columbus's son Fernando and by Bartolemé de las Casas, a Spanish priest who wrote a history of the New World. Both tell us that the learned men at Salamanca were aware of the current debates about the size of the earth, the likelihood of inhabitants in other parts of the world, and the possibility of sailing through the torrid zone at the equator. They challenged Columbus on his claim to having knowledge superior to that of the ancients and on his ability to do what he proposed. They did not, however, deny that the earth was spherical, but rather used its sphericity in their arguments against Columbus, arguing that the round earth was larger than

Columbus claimed and that his circumnavigation would take too long to complete.[14]

When Peter Martyr praised the achievements of Columbus in his laudatory preface to *Decades of the New World* (1511), he was quick to point out that Columbus had proven the equator was passable and that there were indeed peoples and lands in those parts of the globe once thought to have been covered with water. Nowhere, however, did he mention proving the sphericity of the earth.[15] If Columbus had indeed proved the point to doubting scholars, Peter Martyr surely would have mentioned it.

Those who want to preserve Columbus as an icon for the historic moment when the world became round might appeal to the common people. After all, weren't Columbus's sailors afraid of falling off the end of the earth? No, they weren't. According to Columbus's diary, the sailors had two specific complaints. First, they expressed concern that the voyage was taking longer than Columbus had promised. Second, they were frightened that, because the wind seemed to blow constantly due west, they would be unable to make the return voyage eastward.[16]

As we have seen, there is virtually no historical evidence to support the myth of a medieval flat earth. Christian clerics neither suppressed the truth nor stifled debate on this subject. A good son of the church who believed his work was revealing God's plan, Columbus didn't prove the earth was round—he stumbled on a continent that happened to be in his way.

THAT MEDIEVAL ISLAMIC CULTURE
WAS INHOSPITABLE TO SCIENCE

Syed Nomanul Haq

The pious Muslim . . . was expected to avoid . . . [rational]
sciences with great care because they were considered dangerous
to his faith. . . . The *'ulūm al-awā'il* [sciences of the (non-
Muslim) ancients] are pointedly described as "wisdom mixed
with unbelief." . . . They can only lead in the end to unbelief
and, in particular, to . . . the stripping away of all positive
content from God.

> —Ignaz Goldziher, "Stellung der alten islamischen
> Orthodoxie zu den antiken Wissenschaften" (1916)

. . . possession of all of this [Greek] "enlightenment" did not
prompt much intellectual progress within Islam, let alone
eventuate in Islamic science. . . . The result was to freeze
Islamic learning and stifle all possibility of the rise of an Islamic
science, and for the same reasons that Greek learning stagnated
of itself: fundamental assumptions antithetical to science.

> —Rodney Stark, *For the Glory of God* (2003)

Alas, Islam turned against science in the twelfth century. The
most influential was the philosopher Abu Hamid al-Ghazzali
who argued . . . against the very idea of laws of nature, on the
ground that any such laws would put God's hands in chains. . . .
The consequences are hideous.

> —Steven Weinberg, "A Deadly Certitude" (2007)

Between the eighth and fifteenth centuries Islamic culture saw its heyday. At the beginning the followers of the Prophet Muhammad (ca. 570–632), born in the Arabian Peninsula, pushed rapidly across North Africa and up through the Iberian peninsula to the west and eastward to Persia. In 762 the Abbasid caliph al-Mansūr began construction of a new capital, Baghdad, along the banks of the Tigris River in present-day Iraq. By the early tenth century it had become the largest city in the world, with a population of over one million; Cordoba, in Muslim Spain, ranked second.[1] Among the cultural institutions of Baghdad was the House of Wisdom, established as an administrative bureau and library in early Abbasid times. Over the centuries it served as an enduring imperial center for the promotion of scientific activity that fulfilled the consciously forged Abbasid ambition of rivaling the glory of the conquered Persian empire. In this milieu began a massive translation movement to render into Arabic first Sanskrit and Persian texts and then more extensively Greek texts. This development, claims the historian Dimitri Gutas, "demonstrated for the first time in history that scientific and philosophical thought are international, not bound to a specific language or culture."[2]

By the twelfth century grateful scholars in Christian Europe were eagerly translating Arabic scientific texts into Latin—and acknowledging Islamic leadership in natural philosophy.[3] Indeed, even after the original Greek sources became available, some Latin translators preferred using Arabic versions because of the numerous commentaries added by the Muslim sages, who often challenged and corrected the ancient authorities. Nevertheless, denigrators of the Islamic achievement have tended to credit the ancient Greeks with all that was noteworthy in Arabic science, to insist that the Islamic contributors to science were marginal to mainstream Muslim society, and to argue that all scientific creativity had ended by the twelfth century, a fate allegedly caused by opposition from "orthodox" religious leaders such as the philosopher and theologian Abu Hamid al-Ghazālī (1058–1111). I will address each of these claims in turn.

The movement to translate Greek texts into Arabic began in earnest in the ninth century, during the Abbasid dynasty. In terms of "intensity, scope, concentration, and concertedness," claims the historian A. I. Sabra, "it had no precedent in the history of the Middle East or *of the world*."[4] The entire Abbasid elite—soldiers and rulers, merchants and scholars, civil servants and scientists, caliphs and princes—actively supported it with funds and blessings, with patronage cutting "across all lines of religious, sectarian, ethnic, tribal and linguistic demarcation," and including "Arabs and non-Arabs, Muslims and non-Muslims, Sunnīs and Shī'ites." It was, in the words of Dimitri Gutas, an "astounding achievement" with profound consequences for world civilization: "It is equal in significance to, and belongs to the same narrative as, that of Pericles' Athens, the Italian Renaissance, or the scientific revolution of the sixteenth and seventeenth centuries, and it deserves to be so recognized and embedded in our historical consciousness."[5]

Nearly a century ago the influential French scholar Pierre Duhem argued that the Muslim translators "were always the more or less faithful disciples of the Greeks, but were themselves destitute of all originality."[6] But this claim utterly overlooks the historical complexity of transmitting knowledge between cultures. The process of translation involved selection, interpretation, reconfiguration, and transformation; it was a *creative act*. For example, when Qustā ibn Lūqā (820–912) translated Diophantus's Greek *Arithmetica* into Arabic as *The Art of Algebra*, he recast the mathematical operations of the Greek text in terms of a new discipline whose foundations had been laid a little earlier by Muhammad ibn Mūsā al-Khwārizmī (ca. 780–ca. 850); and this marks a fundamental conceptual shift. Similarly, when Abbasid scholars translated Aristotle's *Prior Analytics* as the *Book of Qiyās*, they adopted the Arabic word (taken from the religious sciences) meaning "analogy," which subsequently became the philosopher's term for syllogism. The point is that one cannot in general recover the Greek text by means of reverse translation. To reconstruct a Greek source from an Arabic text, one must move beyond the text and enter into the specific cultural and intellectual setting in which it was created. Clearly, the translation movement, which lasted for over two hundred years, did not just passively preserve the Greek legacy.[7]

According to one recent assessment, "Islamic civilization remained the world leader in virtually every field of science from at least A.D. 800–1300." During this period, writes Sabra, "astronomy tended to be favored as the pursuit most worthy of the attention of both the patron rulers and the patronized mathematicians who were keen on mastering and exploiting the Greek legacy." The Muslims prized astronomy not only for helping to improve astrological predictions and the determination of prayer times, and for demonstrating God's wisdom and perfection, but also for its promise of providing a naturalistic explanation of cosmic phenomena. Spurred largely by a desire to improve their knowledge of the heavens, Muslim astronomers established observatories throughout the region, beginning with one in Baghdad in 828.[8]

The most impressive of these observatories was established in 1259 at Maraghah in a fertile region near the Caspian Sea. Equipped with precision instruments, it flourished under the direction of the Persian Shī'ite astronomer and theologian Nasīr al-Dīn al-Tūsī, who proposed non-Ptolemaic models for the apparent motions of the moon, Venus, and the three superior planets. Adhering to an uncompromising principle of Aristotelian natural philosophy, he succeeded (where the ancient Alexandrian astronomer Ptolemy had failed) in explaining the motions of the planets exclusively in terms of uniform circular motions. During the next century Ibn al-Shātir, a Syrian astronomer who worked as a timekeeper (*muwaqqit*) for ritual prayers in a Damascus mosque, proposed a lunar model that the Polish astronomer Nicolaus Copernicus utilized in *De Revolutionibus* (1543). Indeed, both the Arab timekeeper and the revolutionary Pole used many of al-Tūsī's original and highly sophisticated mathematical techniques.[9]

Medieval Muslims also excelled in medicine. The best-known practitioner was the prolific Persian physician-philosopher Ibn Sīnā, who was active in the early eleventh century and known in the Latin West as Avicenna. His most famous medical work, *The Canon*, sought to bring all medical knowledge, ancient and contemporary, together into one encyclopedic whole. Translated into Latin, it became a staple of medical education in Europe for centuries. Less celebrated but equally important was the thirteenth-century jurist, theologian, and physician Ibn al-Nafis, who effectively discovered the pulmonary circulation of blood—three centuries before its rediscovery by Europeans. A Syrian by birth, Ibn al-Nafis studied medicine in Damascus but spent much of his adult life in Cairo, where, like Ibn Sīnā, he assembled a massive encyclopedia of medicine. Not surprisingly, some contemporaries referred to him as "the second Ibn Sīnā" while others ranked him first.[10]

One of the—if not *the*—leading Islamic men of science was the great tenth-century polymath Ibn al-Haytham, or Alhazen, as Europeans called him. According to the historian David C. Lindberg, he was "the most significant figure in the history of optics between antiquity and the seventeenth century." A highly skilled geometrician, al-Haytham also made important contributions to the development of scientific methodology, especially his linking of mathematics and physics, which the ancient Greeks had pursued separately. He also helped to establish experiment as a category of scientific proof, along with logical demonstration. For this, a writer in the *New York Times*, perhaps indulging in a little myth-making, credited him with conceiving "the greatest idea of the last 1,000 years."[11]

These few examples alone, selected from scores of possibilities, give the lie to assertions that medieval Islam contributed nothing original to science. But what happened in the twelfth century when, as Steven Weinberg has phrased it, "Islam turned against science"? As Weinberg explains it, Muslims fell under the retrogressive influence of "the philosopher Abu Hamid al-Ghazzali who argued . . . against the very idea of laws of nature, on the ground that any such laws would put God's hands in chains." The fountainhead of this myth, it seems, is an erudite Arabist of an earlier generation, Ignaz Goldziher, also quoted at the beginning of this essay. Goldziher, whose historically awkward but ideologically satisfying ideas seem to have reached Weinberg directly or indirectly, emphasized what he considered the negative role of al-Ghazālī, who, we are simplistically told, opposed Hellenistic science—and the very notion of laws of nature—in a book called the *Incoherence of the Philosophers*. Goldziher created the impression that Ghazālī, instead of seeking natural explanations in the manner of the ancient Greeks and their Islamic followers, stressed the unpredictable role played by God and angels. According to Goldziher, his influence helped to bring Islamic science to a screeching halt.

There are several glaring problems with this explanation, not least of which are the examples given above of impressive activity continuing past the twelfth century in astronomy and medicine. Even Goldziher conceded that Ghazālī supported the study of logic and mathematics, but he failed to point out that the allegedly antiscientific Sufi mystic encouraged the pursuit of anatomy and medicine, lamented that Muslims were not doing enough in these sciences, and wrote on anatomy himself. Indeed, the Oxford historian Emily Savage-Smith tells us that Ghazālī's writings served as a powerful spur to the medical sciences.[12]

Goldziher assumed the existence of a dominant Islamic orthodoxy, but what is this thing called "Islamic orthodoxy"? Unlike, say, the Vatican in the Roman Catholic Church, which might promulgate an "official" truth and legislate it by virtue of the institution's coercive powers, Islam has never had a centralized authority. In the Muslim world there are no ordained clergy; no institutionalized religious orders; no synods; and no pontifical truth, a deviation from which would constitute heresy. "At most what one could claim is the prevalence of a certain religious approach at a specific time and specific locality," explains Dimitri

Gutas. "But even this has to be qualified by stating to *whom*, among the different strata of society, this approach belonged, because an assumption of 'prevalence' as meaning 'majority view' is not necessarily always true."[13] Thus it makes no sense to say that Islamic "orthodoxy" turned its back on science. In medieval Islamic society there was an "open marketplace" of ideas, in which some individuals severely criticized natural philosophy in the Greek tradition while others did not.[14]

During the thirteenth and fourteenth centuries, political Islam suffered several severe reversals. In the West, Christians reconquered Spain, taking Cordoba in 1236 and Seville in 1248. From the East, the Mongol Hulagu Khan, a grandson of the notorious Genghis Khan, invaded the heartland of the Islamic world, savagely destroying Baghdad in 1258 and capturing Damascus two years later. The loss of two of its leading intellectual centers, coming on the heels of Ghazālī's critique, might have brought Islamic scientific activity to an end. But, as George Saliba, professor of Arabic and Islamic science at Columbia University, has recently shown, this did not happen. "If we only look at the surviving scientific documents, we can clearly delineate a very flourishing activity in almost every scientific discipline in the centuries following Ghazali," he writes. "Whether it was in mechanics . . . or in logic, mathematics, and astronomy . . . or in optics . . . or in pharmacology . . . or in medicine . . . every one of those fields witnessed a genuine original and revolutionary production that took place well after the death of Ghazali and his attack on the philosophers, and at times well inside the religious institutions." Even "Hulagu's devastating blow" did not prevent Islamic astronomy from experiencing a subsequent "golden age."[15]

By the time of the so-called scientific revolution in western Europe, Islam's scientific star had set in the Middle East, though it continued to shine from a different region of the world within the constellation of Europe. But for centuries, while science in the Latin West had lain in the doldrums, no culture in the world provided a more hospitable home to science than Islam. And no group of Muslims cultivated science more than the religious—and not just in the sense of practicing Islam. As Saliba has pointed out, almost to a man the leading Islamic men of science in the post-Ghazālī centuries "also held official religious positions such as judges, time keepers, and free jurists who delivered their own juridical opinions. Some of them wrote extensively on religious subjects as well, and were more famous for their religious writings than their scientific ones."[16] In other words, they were not inhospitable to science.

THAT THE MEDIEVAL CHURCH PROHIBITED

HUMAN DISSECTION

Katharine Park

From the outset [the sixteenth-century anatomist] Vesalius
proved himself a master. In the search for real knowledge he
risked the most terrible dangers, and especially the charge
of sacrilege, founded upon the teachings of the Church for
ages . . . Through this sacred conventionalism Vesalius broke
without fear; despite ecclesiastical censure, great opposition in
his own profession, and popular fury, he studied his science
by the only method that could give useful results.

> —Andrew Dickson White, *A History of the Warfare of
> Science with Theology in Christendom* (1896)

Pope Boniface VII [sic] banned the practice of cadaver dissection
in the 1200s. This stopped the practice for over 300 years and
greatly slowed the accumulation of education regarding human
anatomy. Finally, in the 1500s, Michael Servetus used cadaver
dissection to study blood circulation. He was tried and
imprisoned by the Catholic Church.

> —Senator Arlen Specter, speaking in favor of S. 2754, the
> Alternative Pluripotent Stem Cell Research Enhancement
> Act (2006)

The myth that the medieval church prohibited human dissection
has several variants. The most basic version, like a number of the
other myths in this book, was largely the creation of Andrew
Dickson White in the late nineteenth century. According to this

version, presented in White's *A History of the Warfare of Science with Theology* and quoted above, Western Christianity was implacably hostile to the study of anatomy through dissection. This attitude was codified by Pope Boniface VIII in his bull *Detestande feritatis* (Of detestable cruelty) of 1299–1300, which threatened those who practiced it with excommunication and persecution. White attributed the church's putative hostility to its commitment to the sacredness of the human body, the divinely created "temple of the soul," and to its doctrine that all human bodies would be resurrected at the Last Judgment. He also cited a supposed conciliar decree of 1248, *Ecclesia abhorret a sanguine* (The church abhors the shedding of blood), which prohibited the practice of surgery by monks and priests.[1]

More recent versions of the myth acknowledge that dissection was in fact both prescribed and practiced in a number of late-medieval universities. They emphasize, however, that dissection was limited to the bodies of executed criminals, who had forfeited any claim to reverence or salvation. In either case, the story goes, the sacrilegious nature of dissection, combined with popular superstition, meant that it was rarely practiced. This version ascribes a slavish adherence to bookish authority on the part of medieval anatomists: those few courageous souls who were committed to the rational study of the human frame on the basis of direct experience—most notably Leonardo da Vinci (1452–1519); Andreas Vesalius (1514–1564), author of the famous illustrated anatomical textbook *On the Fabric of the Human Body* (1543); and, in Senator Arlen Specter's mind, the Spanish physician and theologian Michael Servetus (1511–1553)—were forced to snatch cadavers from tombs and gibbets and dissect them clandestinely in the dead of night.

Although there are just enough points of contact between this story and historical reality to give it a veneer of plausibility, the situation was in fact considerably more complex. Most medieval church authorities not only tolerated but encouraged the opening and dismemberment of human corpses to religious ends: the embalming of holy bodies by evisceration; their division to yield corporeal relics; the inspection of the internal organs of holy men and women for signs of sanctity; and the operation that came later to be known as caesarean section, whose aim was to baptize fetuses extracted from the bodies of women who died in childbirth. All of these practices give the lie to the claim that the church as an institution was committed to the integrity of the human body after death, as does the widespread practice of dividing the corpses of princes and nobles before burial. At the same time, medieval culture placed distinct limits on the acceptable treatment of human corpses, which dramatically restricted the number of cadavers available for dissection. But these limits reflected secular values of personal and family honor and ritual decorum and were enforced by local governments rather than by religious authorities.[2]

The facts, then, are as follows: Human dissection does not seem to have been practiced with any regularity before the end of the thirteenth century in either pagan, Jewish, Christian, or Muslim cultures. The only exception was a brief period in the fourth to third century B.C.E., when Herophilus and Erasistratus, two Greek medical scholars working in the Egyptian city of Alexandria, made a series of studies of the human body based on dissection.[3] While the Greek and Roman avoidance of human dissection seems to have had roots in the belief that corpses were ritually unclean, early Christian culture broke definitively with the idea of corpse pollution, embracing tombs as holy places and the bodies of the dead as objects of veneration and potential sources of magical and healing power.[4] Although the church did not prohibit human dissection in the early Middle Ages, there is no evidence of its practice. This may have reflected in part the disapproval of early Christian writers such as Augustine (354–430), who saw the fascination with dismembered corpses not as sacrilegious but as a kind of unhealthy curiosity regarding matters irrelevant to salvation. But it had at least as much to do with the generally undeveloped state of medical learning in western

Europe after the fall of the western Roman Empire in the fifth century, where medical teaching and research of all sorts reached a low ebb.[5]

In the late thirteenth century we find the first evidence of the opening of human bodies on the part of medical men, in connection with municipally mandated autopsies to determine cause of death in the interests of criminal justice or public health. The appearance of human dissection—the opening of corpses in the service of medical teaching and research, continuous with modern academic practices—took place around 1300 in the Italian city of Bologna, home of what was arguably the greatest medical faculty of the day. Inspired by renewed interest in the works of the Greek medical writer Galen (ca. 129–ca. 200) and his Arabic followers, none of whom are known to have dissected human beings, medical teachers and students at Bologna began to open human bodies, and Mondino de' Liuzzi (ca. 1275–1326) produced the first known anatomy textbook based on human dissection, which remained a staple of university medical instruction through the early sixteenth century. Initially, dissection was confined to Italian universities and colleges of physicians or surgeons, a number of which adopted it as an annual requirement, and to the southern French university of Montpellier. In the late fifteenth century, however, the practice spread to medical faculties in northern Europe, and by the sixteenth century it was widely performed in universities and medical colleges in both Catholic and Protestant areas.[6]

The official practice of human dissection in late medieval Italian universities obviously calls into question the two official ecclesiastical prohibitions of dissection cited by White and other proponents of the myth. The first such prohibition, *The church abhors the shedding of blood,* which White described as promulgated at the Council of Le Mans in 1248, was shown forty years ago to be a "literary ghost," produced by an inept eighteenth-century French historian. While there were in fact several decrees forbidding clerics in higher orders from practicing any form of surgery involving cautery (burning) or cutting, these reflected concern that clergy might be putting people's lives in danger for pecuniary reasons and had nothing whatsoever to do with dissection.[7] The history of White's second purported prohibition, the bull of Pope Boniface VIII called *Of detestable cruelty,* raises more complicated issues. First promulgated in 1299, it forbade under pain of excommunication a contemporary funerary practice that involved cutting up the corpse and boiling the flesh off the bones, in order to make it easier to transport for distant burial—a procedure that had gained currency among European crusaders to the Holy Land. Although there is no evidence that Boniface had the practice of dissection in mind when issuing this bull, it nonetheless had an indirect impact on the study of anatomy in fourteenth- and fifteenth-century Europe. This impact was negligible in Italy, where it was taken literally; Mondino, for example, noted only that it prevented him from boiling the bones in the ear to make them easier to examine. But the bull seems to have been understood more expansively by some anatomists active in northern Europe, who interpreted it either as categorically forbidding dissection or as forbidding dissection without a papal dispensation. Others, such as the great fourteenth-century French surgeon Guy de Chauliac (ca. 1290–ca. 1367/70), showed no hesitation about dissection at all.[8] Crucially, avoidance of dissection seems to have reflected preemptive caution on the part of anatomists rather than actual ecclesiastical pressure: I know of no case in which an anatomist was ever prosecuted for dissecting a human cadaver and no case in which the church ever rejected a request for a dispensation to dissect. Certainly, there is no convincing evidence that Vesalius ran afoul of church authorities two hundred years later. White's contention that he suffered "ecclesiastical censure" is based on a single, highly dubious source, a letter that appears to be either a piece of sixteenth-century anti-Catholic polemic or a seventeenth-century Protestant fabrication. In any case, this letter has nothing to do with dissection proper: it attributed Vesalius's purported denunciation by the Inquisition to the fact that he had autopsied a patient who turned out not to have been dead after all.[9] And although Michael Servetus was in fact tried and executed in the mid-sixteenth century, his condemnation was purely theological and had nothing to do with his medical activities.

Unlike dissection, however, grave robbing was clearly and repeatedly prohibited by both secular and ecclesiastical authorities, and it was grave robbing that got aspiring anatomists into trouble, as in the case of four medical students at Bologna who were prosecuted in 1319—by the city, not the church—for exhuming and dissecting the corpse of a criminal who had been executed earlier that day.[10] From at least the early fourteenth century, the study of anatomy was hampered by a chronic shortage of cadavers, which led enterprising students to violate graves and even, in the sixteenth century, to hijack corpses from funeral processions. This shortage resulted not from religious prohibitions but from strong cultural sensitivities concerning who was and was not an appropriate subject for dissection, sensitivities that were enshrined in local law. The problem with dissection from the point of view of the inhabitants of late-medieval and Renaissance Europe was not that it was sacrilegious but that it was a gross dishonor to the individual and, more to the point, to his or her family. To be exhibited naked in front of a group of university students—augmented, in the sixteenth century, by local notables and visiting dignitaries—was a deeply shaming prospect, particularly since dissection rendered the body unsuited for a normal funeral, where the corpse was usually transported on an open bier. Conversely, families had no qualms about autopsies, which were becoming increasingly common in this period, since they were performed privately and left the body substantially intact for the funeral procession.[11]

Municipalities responded to these concerns by restricting dissection to the bodies of foreigners, often defined as those born more than thirty miles away, who presumably had no family members on the scene to be dishonored. In the case of what were known as "public" dissections, ceremonial teaching events carried out once or twice a year, this foreign cadaver was almost always that of a criminal who had just been executed—an administratively economical arrangement that allowed the city to monitor the provenance of the corpse and minimized the possibilities for foul play. In the years after 1500, however, anatomy exploded as a research field, creating a demand for cadavers that the meager trickle of bodies from the gallows could not satisfy. Anatomists looked increasingly to the other obvious source of foreign corpses: people who had died in local hospitals, which took in the sick, elderly, and disabled who had no families to care for them and who were therefore ideal candidates for dissection. The best-known example of this practice was Leonardo's dissection of an aged patient in the great Florentine hospital of Santa Maria Nuova (a religious institution), whom he had befriended and attended on his deathbed.[12]

The myth of church prohibition of human dissection is as strong now as when it was first invented by White in the late nineteenth century, although its focus has shifted in interesting ways. For one thing, the heroic figures presented as braving the putative prohibition are more likely now to be artists rather than scientists, as in White's version; Italian tour guides regularly regale visitors to Florence and Milan with stories about Leonardo's anatomical researches in church basements with secret exits and his invention of mirror writing to hide his results. In fact, Leonardo, like Vesalius (and Servetus), never got into trouble for practicing human dissection, though his difficulty in obtaining corpses forced him to rely largely on animal cadavers. This difficulty related not to religious restrictions, however, but to the fact that he was an artist, with no medical training and no institutional credentials to legitimize his work.

THAT COPERNICANISM DEMOTED HUMANS
FROM THE CENTER OF THE COSMOS

Dennis R. Danielson

Copernicus dethroned the Earth from the privileged position that
Ptolemy's cosmology accorded it.

—Sir Martin Rees, *Before the Beginning* (1998)

Dethronement of Earth from the centre of the universe caused
profound shock: the Copernican system challenged the entire
system of ancient authority and required a complete change in
the philosophical conception of the universe.

—*Britannica Concise Encyclopaedia* (2007)

Nearly a century ago the psychoanalyst Sigmund Freud (1856–
1939) alleged that science had inflicted on humanity "two great
outrages upon its naïve self-love": the first, associated with the
sixteenth-century astronomer Nicolaus Copernicus (1473–1543),
"when it realized that our earth was not the centre of the universe,
but only a tiny speck in a world-system of a magnitude hardly
conceivable"; the second, when Charles Darwin (1809–1882)
showed that humans had descended from animals. (Freud thought
that his own psychoanalytic theory, which showed that humans
act under the influence of unconscious urges, delivered "the third
and most bitter blow.") Freud's observation was not new. Already
for more than a century, the Polish astronomer was commonly
portrayed as a kind of killjoy as far as humankind and its cosmic

value are concerned. It seems neither popularizers nor serious scientists can utter Copernicus's name without at once feeling the urge to say that he "dethroned" the earth or us humans when he explained that the earth circles the sun, instead of the other way around. Almost every week new examples of the same claim appear in newspapers, on the web, and on the syllabuses of college courses—it is repeated so often, and by such respectable voices, that it has become like a perennial mold in our collective mental cupboards and a gratuitous blight on planetary morale.[1]

This great Copernican cliché simplistically assumes that central is good, or special, and that to be removed from the center is bad.[2] Sometimes, of course, such claims are justified, though often by *central* we mean figurative location, not literal, as in "My spouse and children stand at the center of my affections." Other times, whether centrality implies specialness just depends. For example, whether the center of a particular city is a special or desirable location depends on the city. In some cases it might be, and in others it's not. To judge whether Copernicanism did actually constitute a demotion of earth, we therefore require a proper "market evaluation" of earth's immediate neighborhood at the center of the universe when, in the sixteenth century, Copernicus relocated it to what were then regarded as the cosmic suburbs.

A further, overlapping assumption of the cliché is that geocentrism equates with anthropocentrism. The Ptolemaic system of astronomy, which Copernicanism eventually superseded, was indeed geocentric in the sense that it placed earth *(geo-)* at the center of the universe literally and geometrically—for reasons that will become apparent. However, *anthropocentrism* is a figurative term only. Like egocentrism or ethnocentrism, it signals a person's or a group's attitude to the value of something, in this case of humankind *(anthropos)*. The first time I ever visited London, England, I was given a tour by a proud Londoner who pointed out Piccadilly Circus by announcing: "And *that* is the center of the universe." He was speaking figuratively, with conscious irony, about the *importance* of the place. If on the basis of that comment I had challenged my guide's cosmology, he would rightly have thought me confused as well as humorless and condescending. For similar reasons, anthropocentrism and geocentrism shouldn't be treated as if they were the same thing.

The other presumption that generally piggybacks on the cliché is that Copernicus, in allegedly reducing the status of the earth, also struck a blow against religion, particularly the Abrahamic religions, which supposedly require the cosmic centrality of humankind. The weaknesses of this view, as already hinted, include a failure to distinguish figurative from literal centrality: If any of the great religions requires humankind to be central, it is surely only in the figurative sense of its value, not in the literal sense of its location. Yet even that is a big "if." The Jewish and Christian scriptures, for example, do not promote "naïve self-love" (Freud's term) on the part of humans, but instead proclaim our smallness, weakness, and often moral incapacity against the immense greatness, goodness, and otherness of the Creator. In the biblical tradition, it is this awareness that frames one's sense of privilege at being an object of divine blessing *despite* one's conspicuous finitude and smallness relative to the world at large—as expressed in the Psalter: "When I look at your [God's] heavens, the work of your fingers, the moon and the stars that you have established; what are human beings that you are mindful of them?" (8:3–4; NRSV).

It is true of course that in the seventeenth century the arch-Copernican Galileo Galilei (1564–1642) met opposition from Catholic authorities in Rome. However, their dispute focused on matters related to biblical interpretation, educational jurisdiction, and the threat Galileo represented to the entrenched "scientific" authority of Aristotle, not on any supposed Copernican depreciation of the cosmic specialness or privilege of humankind. If anything, Galileo and his fellow Copernicans were *raising* the status of earth and its inhabitants within the universe.

According to Aristotle (384–322 B.C.), whose system of physics dominated European natural philosophy from his own time on into the seventeenth century, earth stood at the center of the universe because that is where heavy things naturally collected. On this view, it is the universal centerpoint, not a massive body, that draws heavy things to itself. Aristotle taught that place itself "exerts a certain influence"[3] and it is merely the fact that the earth is composed of the heaviest stuff (earth being heavier than the other three elements: water, air, and fire) that explains why the body on which we live must be motionless *at* the center of the universe.

Furthermore, earth's central position was taken as evidence not of its importance but (to use a term still in circulation) its grossness. The great medieval Jewish philosopher Moses Maimonides (1135–1204) asserted that "in the case of the Universe . . . the nearer the parts are to the centre, the greater is their turbidness, their solidity, their inertness, their dimness and darkness, because they are further away from the loftiest element, from the source of light and brightness." Similarly, Thomas Aquinas (1225–1274), the foremost Christian philosopher of the Middle Ages, declared that "in the universe, earth—that all the spheres encircle and that, as for place, lies in the center—is the most material and coarsest *(ignobilissima)* of all bodies."[4] Making a consistent extrapolation from this view, Dante, writing his *Inferno* in the early fourteenth century, placed the lowest pit of hell at the very midpoint of the earth, the dead center of the whole universe.

Pre-Copernican cosmology thus implied not the figurative, metaphysical centrality—the importance or specialness—of earth, but rather its physical centrality and at the same time sheer grossness. This negative view extended even beyond the Middle Ages. In 1486, in a work often considered a manifesto of the Italian Renaissance, Giovanni Pico referred to earth as occupying "the excrementary and filthy parts of the lower world." And in 1568, a quarter-century after the publication of Copernicus's *Revolutions,* Michel de Montaigne asserted that we are "lodged here in the dirt and filth of the world, nailed and rivetted to the worst and deadest part of the universe, in the lowest story of the house, and most remote from the heavenly arch."[5]

While rightly anticipating some opposition to his theory of a planetary earth, Copernicus retorted that he would "not waste time" on "idle talkers" who cited the Bible's apparent references to a fixed earth and moving sun. The Christian tradition, at least among the learned, had ample interpretive means for distinguishing between what science teaches and how people commonly talk (as, for example, we still do when referring to sunrise and sunset).[6] The greater obstacle to Copernican theory was almost two millennia of a physics according to which the gross earth was obviously down and the glorious sun obviously up. Accordingly, the first semiofficial response to the *Revolutions*—written by a Dominican friar but framed in transparently Aristotelian terms—complained that "Copernicus puts the indestructible sun in a place subject to destruction."[7] Rather scandalously, heliocentrism was seen as "exalting" the position of humankind in the universe and pulling the earth up out of the cosmic sump that Copernicus's predecessors thought it occupied—and conversely, placing the divinely associated sun into that central yet tainted location.

To preempt this charge, Copernicus and his followers did what they could, rhetorically, to renovate the cosmic basement. Copernicus's disciple Rheticus (1514–1574) offered a governmental analogy: "My teacher . . . is aware that in human affairs the emperor need not himself hurry from city to city in order to perform the duty imposed on him by God." The centrality and immobility of the sun in Copernicus's system were therefore perfectly consistent with—indeed were essential to—the sun's dignity and its efficient governance of the planets. Continuing this theme of orderly rule, Copernicus tried to enhance the status of the center by envisaging it as an advantageously located throne *(solium)* that formed a poetically fitting place from which the kingly sun *(sol)* could illuminate and govern his subjects. In Copernicus's cosmology, the center was transformed into a place of honor, while at the same time earth was promoted to the status of a "star" that "moves among the planets as one of them."[8]

It was a remarkable feat: Copernicus thus simultaneously enhanced the cosmic status of both earth and sun. And the latter part of this task succeeded so well that, ever since, the earth's removal from what became the sun's place of honor has appeared to some as a diminution of its value. A refutation of that anachronistic interpretation, however, can be put quite simply: For earth to be raised up out of what were then considered "the excrementary and filthy parts of the lower world" can't seriously be interpreted as a demotion.

The seventeenth century's foremost Copernicans expressed exhilaration at earth's release from the dead center of the universe. In 1610 Galileo explicitly presented his account of earthshine—of how sunlight reflected by the earth's surface illuminates the moon, just as moonlight does the earth—as entailing "commerce" between these two heavenly bodies: "The earth, with fair and grateful exchange, pays back to the moon an illumination like that which it receives from the moon." Moreover, Galileo saw this account as militating against "those who assert, principally on the grounds that it has neither motion nor light, that the earth must be excluded from the dance of the stars. For . . . the earth does have motion, . . . it surpasses the moon in brightness, and . . . it is not the sump where the universe's filth and ephemera collect."[9] Whereas Ptolemaic cosmology implied that the place of earth was both low and lowly, Galileo could see that humanity's new Copernican perspective was, in more senses than one, *uppity*.

Johannes Kepler (1571–1630) likewise saw earth's new planetary position as affording humankind great cosmic advantage. He argued that, because "man" had been created for contemplation,

he could not remain at rest in the center . . . [but] must make an annual journey on this boat, which is our earth, to perform his observations . . . There is no globe nobler or more suitable for man than the earth. For, in the first place, it is *exactly in the middle* of the principal globes . . . Above it are Mars, Jupiter, and Saturn. Within the embrace of its orbit run Venus and Mercury, while at the center the sun rotates.

To actualize their divine image properly, humans must be able to observe the universe from a (redefined) "central" but dynamic and changing point of view conveniently provided by what Kepler saw as our optimally placed orbiting space station called earth. Only with the *abolition* of geocentrism, then, might we truly say that we occupy the best, most privileged place in the universe.[10]

Other Copernicans also saw earth's "relocation" in positive terms. In England, the foremost proponent of the new astronomy was a clergyman named John Wilkins (1614–1672). Wilkins openly opposed those who presumed that "the earth is of a more ignoble substance than any other planets, consisting of a more base and vile matter," that "the centre is the worst place," and therefore that the center is where earth must be located. Especially with the decline of Aristotelian physics by the mid-seventeenth century, Wilkins could denigrate these claims as "(if not evidently false) yet very uncertain."[11]

It is hard to know exactly what gave rise to the myth of religious opposition to the supposed "dethroning" of the earth, apart from Copernicus's apparent success in rebranding the cosmic dead center as a royal throne for the sun. Other kinds of opposition that were not religious in nature might have blurred with the question of the uniqueness of earth once it was seen as one planet among others. A particular scientific difficulty was that, until the work of Isaac Newton late in the seventeenth century, there was no new physics to explain how something as heavy as the earth could pursue a stable orbit about the sun. Copernicanism also demanded that the universe be significant orders of magnitude larger than anyone had previously imagined. Thus for many, undoubtedly, the exhilaration of a Galileo or a Kepler could translate instead into bewilderment. One thinks of John Donne's oft-quoted lament (1611), "'Tis all in pieces, all coherence gone," or Blaise Pascal's "The eternal silence of these infinite spaces frightens me" (1670). Yet the increased size and awesomeness of the universe could just as readily inspire increased, not diminished, religious awe and reverence. American natural philosopher and Puritan clergyman Cotton Mather (1663–1728), writing about the astounding cosmic vistas opened up by the telescope, exclaimed, "Great God, what a variety of worlds hast thou created! How astonishing are the dimensions of them! How stupendous are the displays of thy greatness, and of thy glory!"[12]

The Copernican cliché seems to have appeared for the first time in France more than a century after the death of Copernicus as part of an antianthropocentric critique. Cyrano de Bergerac (1619–1655) associated pre-Copernican geocentrism with "the insupportable arrogance of Mankinde, which fancies, that Nature was onely created to serve it." Most influentially, the science popularizer Bernard le Bouvier de Fontenelle's *Discourse of the Plurality of Worlds* (1686) complimented Copernicus—who "takes the Earth and throws it out of the center of the World"—for his "design was to abate the Vanity of men who had thrust themselves into the chief place of the Universe." This interpretation became the standard and apparently unquestioned version of the Enlightenment, as magisterially summarized in 1810 by Johann Wolfgang Goethe, who repeated the notion that "no discovery or opinion ever produced a greater effect on the human spirit than did the teaching of Copernicus," for it obliged earth "to relinquish the colossal privilege of being the center of the universe."[13]

In the hands of some, the myth of earth's dethronement appears more than a mere anachronism or disinterested misunderstanding. For when Fontenelle and his successors tell the tale, they are openly "very well pleased" with the demotion they read into the accomplishment of Copernicus. But a trick of this supposed dethronement is that, while purportedly rendering "Man" less cosmically and metaphysically important, it actually enthrones us modern "scientific" humans in all our enlightened superiority.

And often it insinuates, without warrant, that scientific advance is inevitably accompanied by an abandonment of the quest—a quest that may encompass what is sometimes called religion—to grasp humankind's possible purpose or significance within the universe as a whole. By equating anthropocentrism with the now plainly untenable geocentrism, such modern ideology dismisses as nugatory or naïve the legitimate and still-open question about the role that earth and its inhabitants may play in the dance of the stars. Instead it offers, if anything at all, a role that is cast in exclusively existential or Promethean terms, with humankind lifting itself up by its own bootstraps and heroically, though in the end pointlessly, defying the universal silence.

Historically and philosophically, however, the tale is a fabrication. Reasonable people need not believe it.

THAT GIORDANO BRUNO WAS THE FIRST MARTYR OF MODERN SCIENCE

Jole Shackelford

> Bruno stood in silence before the fifteen men. Severina read the charges, a total of eight counts of heresy. These included his belief that the transubstantiation of bread into flesh and wine into blood was a falsehood, that the virgin birth was impossible, and, perhaps most terrible of all, that we live in an infinite universe and that innumerable worlds exist upon which creatures like ourselves might thrive and worship their own gods.
>
> —Michael White, *The Pope and the Heretic: The True Story of Giordano Bruno, the Man Who Dared to Defy the Roman Inquisition* (2002)

> In connection with his Copernican beliefs, he held also that the universe contains an infinite number of worlds populated with intelligent beings. On account of these teachings, Bruno was tried for heresy by the Inquisition and burned at the stake in 1600. He thus became the first martyr of modern science at the hands of the Church, and thereby a precursor to Galileo . . . The facts of this myth are true, though sketchy to the point of poverty and generally misleading in their emphasis.
>
> —Edward A. Gosselin and Lawrence S. Lerner, in *The Ash Wednesday Supper: La Cena de le ceneri* (1977)

By sixteenth-century European measures, Giordano Bruno was a heretic. His doubts about virgin birth and the identification of God with Christ, whom he regarded as a clever magician, were repugnant to every major Christian denomination, Protestant and Catholic alike. His refusal to recant these and various other propositions specified in charges brought against him by the Roman Inquisition in the last years of the century led to his conviction and condemnation to death as an unrepentant heretic in January of 1600. On the seventeenth of February he was publicly and ceremoniously burned at the stake, alive, in Rome's Flower Market. His end is brutal to modern sensibility but not exceptional in the early modern period, when traitors and other serious malefactors were drawn and quartered, eviscerated while still living, and their parts displayed on gallows or bridges for all to see. Yet Bruno's death stands out, mentioned in passing in most popular and even academic surveys of the emergence of modern scientific ideas and practices during what has been called the Scientific Revolution of the sixteenth and seventeenth centuries. Indeed, although more careful historians clearly label Bruno a heretic, many also link his heresy to important innovations in scientific cosmology. Specifically, his advocacy of a version of Copernicus's heliocentric planetary hypothesis and the idea that our universe is infinite, with many suns and planets, is regarded as an early imaginative exploration of what became the open universe of René Descartes, Isaac Newton, and Pierre-Simon Laplace. A few go so far as to identify him as the first scientific martyr on account of this linkage—an incendiary example of an inevitable collision between rigid theological dogma and freedom of speculation within natural philosophy, the precursor to modern science.

The sense of this confrontation and the myth of Bruno as martyr for his scientific beliefs was stated concisely in *The Warfare of Science* (1876) by Andrew Dickson White, who, along with his contemporary John William Draper, did much to set the modern tone of the historical conflict between religion and science:

> He [Bruno] was hunted from land to land, until, at last, he turns on his pursuers with fearful invectives. For this he is imprisoned six years, then burned alive and his ashes scattered to the winds. Still the

new truth lived on; it could not be killed. Within ten years after the martyrdom of Bruno, after a world of troubles and persecutions, the truth of the doctrine of Kopernik was established by the telescope of Galileo.[1]

While White did not explicitly say that Bruno was put to death because of his scientific ideas—namely, his promotion of the new heliocentric cosmology of Nicolaus Copernicus—the mythical connection is implicit in his statement: Bruno was a Copernican and he was persecuted and martyred, but the Copernican truth could not be killed with him; Galileo proved this truth soon after his martyrdom. White did not create the myth of Giordano Bruno's scientific martyrdom, but his book placed Bruno into a larger dialogue about the relationship between philosophical freedom and control of religious teaching that exerted a great influence on the modern intellectual history of the West.[2]

This damning equation of Bruno's Copernican cosmology and his fiery extermination by the Holy Office has persisted, as is evident in Hugh Kearney's survey of the Scientific Revolution, *Science and Change 1500–1700* (1971): "Bruno was the most enthusiastic exponent of the heliocentric doctrine in the second half of the century. He lectured throughout Europe on it and in his hands Copernicanism became part of the Hermetic tradition"; "Bruno transformed a mathematical synthesis into a religious doctrine"; and "[i]nevitably these views brought him into conflict with the orthodox academics."[3] Kearney's account is richer and more sophisticated than White's because he recognized that Bruno had used Copernicus's ideas not in a scientific context but in a specifically religious context, namely, the advocacy of Hermetic religion as a corrective for the woes of Reformation and Counter-Reformation Europe. Kearney thus distanced Bruno from astronomy but continued to identify him with the new mathematical hypothesis that "inevitably" conflicted with religious orthodoxy.

Kearney's reading of Bruno as mainly a religious thinker and writer rather than as a natural philosopher or scientist resulted in part from closer scrutiny of the larger scope and context of Bruno's work, forcefully articulated by Frances Yates in *Giordano Bruno and the Hermetic Tradition* (1964), and partly from the identification of Western intellectual history with a positive development toward modern science. According to this positivist view of history, Bruno was not a martyr to science because he was no scientist at all![4] Moreover, inasmuch as the cosmological ideas that Bruno professed were heretical in nature, and Bruno himself was an apostate priest and member of a monastic order and therefore formally under the legal jurisdiction of the Catholic church, the Inquisition was quite within its legal rights to prosecute him. According to Angelo Mercati, who discovered and published the summary document pertaining to Bruno's trial and condemnation by the Roman Inquisition, Bruno's crimes were clearly of a religious nature, no matter what his views of the structure of the physical cosmos.[5]

Despite Mercati's authoritative assessment of the surviving documentary evidence, which has itself been criticized recently as biased, the association of Bruno's cosmology with his condemnation and execution lingers, both in academic articles and in popular treatments of the history of scientific thought.[6] William Bynum, for example, is careful to note the ecclesiastical legal proceedings against Bruno for charges of heresy but links them to his cosmology by juxtaposition:

> Both the influence of Neoplatonism during the Renaissance and the astronomical discoveries during the scientific revolution raised the conceptual and physical possibility that an infinite variety of beings required infinite space in which to exist, that the earth was only one of a large number of inhabited planets . . . Such a notion obsessed the scientific mystic and martyr Giordano Bruno (1548–1600), who died at the stake (though probably for his interest in magic rather than his devotion to plenitude). *Nevertheless, Bruno's fate highlights the fact that ideas of the plurality of worlds were hard to square with the biblical account of man's place in the cosmos and with the Fall and unique Incarnation of Jesus.*[7]

The durability of the association of Bruno the Heretic with an inevitable collision between the authority of Christian dogmatic theology and the freedom of scientific thought is plain from the dust jacket of a recent, somewhat fictionalized popular account of Bruno's life and fiery death, Michael White's *The Pope and the Heretic: The True Story of Giordano Bruno, the Man Who Dared to Defy the Roman Inquisition* (2002). The marketing statement beginning inside the dust jacket promises

> [t]he compelling story of one of history's most intriguing yet little-known natural philosophers—a sixteenth-century Dominican priest whose radical theories influenced some of the greatest thinkers in Western culture—and *the world's first martyr to science* . . . The Inquisition's attempts to obliterate Bruno failed, as his philosophy and influence spread: Galileo, Isaac Newton, Christian Huygens, and Gottfried Leibniz all built upon his ideas . . . a martyr to free thought.[8]

Again we see the implicit reasoning: Bruno was an innovative natural philosopher; he was executed by the church for his ideas, which eventually formed a basis for modern science; ergo, the church killed him to limit the free development of scientific ideas. How did this defrocked monk and unrepentant heretic who denied the doctrine of the Holy Trinity—the key to Catholic teaching of redemption and eternal life—come to be "the world's first martyr to science"?

Part of the answer to this question lies in the aims of nineteenth-century historiography. For various reasons, post-Enlightenment historical essayists sought to exalt Bruno as an exemplary figure in the struggle for free thought against the confining authority of aristocratic government supported by religious orthodoxy. Already in the eighteenth century, writers placed Bruno's cosmology in the canonical development of Western thought, and during the course of the nineteenth his place in the history of science was made secure by such widely read authors as William Whewell, John Tyndall, and Henry

Fairfield Osborn.[9] In the last quarter of the century this propaganda impulse fed the secular and modernist ambitions of Italian unification, visibly evident in the successful international effort to raise a statue to the martyr on the spot of his combustion. As a rebel against state tyranny, Bruno did well during the struggles with totalitarianism that marked much of the twentieth century as well.[10] But the prejudices and opportunism of historiography do little to explain the fact that, already in the seventeenth century, Bruno had become an emblem for freedom of philosophy, especially natural philosophy, embedded in the story of the emergence of the new science by the very people who undertook scientific revolution.

The disproportionate role of Galileo and the triumph of heliocentric cosmology, which is arguably the most dramatic and iconic manifestation of the new science, secured Bruno's position as a martyred, visionary forerunner to Galileo's own struggle to free philosophy from the constraints of dogmatic theology in Catholic Italy. Volker Remmert has cogently asserted that the technology of book production facilitated what he and William B. Ashworth have described as an ideological war being played out in the frontispieces and title-page illustrations to seventeenth-century cosmological treatises, which positioned Galileo in his own time as the "defining emblem of the Copernican Revolution, if not of the whole Scientific Revolution."[11] In a sense, then, it was Galileo who willy-nilly and apparently unwittingly gave Bruno salience in the historical record. But it is Johannes Kepler's criticism of Galileo for failing to credit his heretical countryman as a precursor in promoting Copernican cosmology that most profoundly illustrates that Galileo's contemporaries already thought of Bruno as a scientific thinker, not merely as a religious heretic. This undermines the claim that Bruno's cosmology can be disregarded as the undisciplined ramblings of a speculative philosopher who sought support for a reform of religion and was duly prosecuted for his efforts.[12] In light of Kepler's perception of Bruno as a legitimate participant in natural-philosophical discussion, it is understandable that historians of science would include him in the history of science and frame Galileo's conflict with Cardinal Robert Bellarmine (who presided over Bruno's condemnation) in terms of "the specter of Giordano Bruno." Accordingly, Bruno's burning stood as an example of the Inquisition's hostility to philosophical claims that had serious theological implications for core Catholic doctrine as defined by the Council of Trent.[13] In this regard, Mercati was correct; the church was within its legal right to condemn Bruno for certain philosophical views as well as the more obvious anti-Trinitarian and other theological heresies. But this returns us to the original question: Was Giordano Bruno burned alive because of his acceptance of Copernicanism and advocacy of an infinite universe?

This question is twofold, and it is clear that the answer to the first part is "no." The Catholic church did not impose thought control on astronomers, and even Galileo was free to believe what he wanted about the position and mobility of the earth, so long as he did not *teach* the Copernican hypothesis *as a truth* on which Holy Scripture had no bearing. However, the second part of this question is clouded by the documentation surviving from Bruno's trial, the interrogations of both the Venetian and Roman Inquisitions, which reveals a nagging concern on the part of Bruno's inquisitors about his idea that there may be innumerable, inhabited earth-like worlds in a universe made infinite by God's infinite powers and scope. As a trifle of philosophical speculation, the plurality of worlds was of little consequence to post-Tridentine theologians; as a truth, however, it would undermine the central teaching of the Christian faith. According to biblical theology, humans are unique creations, made in the image of God and placed on an earth that was created *ex nihilo*. Moreover, the New Testament teaches that Christ offered humans redemption from the original sin of Adam; if Bruno were right and there were other worlds, the specter of other creations with other Adams and humans would vitiate the uniqueness of human salvation and the hope of resurrection and eternal life upon which the Christian faith was founded. Whether the persistent questions about Bruno's cosmological propositions figured into the final deliberation and sentencing of the heretic must remain unknown, unless the actual records of Bruno's last months, which apparently disappeared in the aftermath of Napoleon's failed conquest of Europe, are discovered. Yet the fact remains that cosmological matters, notably the plurality of worlds, were an identifiable concern all along and appear in the summary document: Bruno was repeatedly questioned on these matters, and he apparently refused to recant them at the end.[14] So, Bruno probably was burned alive for resolutely maintaining a series of heresies, among which his teaching of the plurality of worlds was prominent but by no means singular. But was this then a scientific question—a matter for philosophers—or a religious doctrine that constituted a serious breach of the letter of scriptural revelation, church discipline, and centuries of Catholic tradition?

Current historians look with relative disdain on nineteenth- and twentieth-century efforts to construe the past in terms of rigidly defined modern categories, such as those that would distinguish Bruno's scientific speculations from his conception of religious truths. We must look beyond the construction of the myth of Giordano Bruno as a moralistic *topos* in the triumphant struggle between the freedom of scientific inquiry and the shackles of conformity to the dead letter of religious revelation. Instead we must examine the actor's own contexts for clues to meanings and categories that can explain his history. In Bruno's day, indeed in his own writings, theology and philosophy were of one piece, inseparable. He stated this succinctly in the prefatory letter dedicating *The Cabala of Pegasus* (1585) to the fictional Bishop of Casamarciano: "I don't know if you are a theologian, philosopher, or cabalist—but I know for sure that you are all of these . . . And therefore, here you have it—cabala, theology and philosophy; I mean, a cabala of theological philosophy, a philosophy of kabbalistic theology, a theology of philosophical cabala."[15] Clearly Bruno thought of his work as all three and incomplete if construed as any one of them alone; he wrote as a philosopher but reckoned himself a Professor of Sacred Theology.[16] As a result, as Mercati pointed out, "The Church could intervene, was bound to intervene and did intervene"—in so doing, it issued a clear warning to those who would wield "Pythagorean" natural philosophy, the name applied to the heliocentric hypothesis by some in church circles, as a weapon against the approved Christian faith.[17] The association of Copernicus's ideas with the ancient central fire cosmology of Pythagoras was more than a dismissal of the antiquity of heliocentrism; it was especially damning, inasmuch as it implied other shared heresies, such as the Pythagorean belief in the transmigration of souls. Such teachings were not to be tolerated in post-Tridentine Rome.

THAT GALILEO WAS IMPRISONED AND TORTURED
FOR ADVOCATING COPERNICANISM

Maurice A. Finocchiaro

[T]he great Galileo, at the age of fourscore, groaned away
his days in the dungeons of the Inquisition, because he had
demonstrated by irrefragable proofs the motion of the earth.

   —Voltaire, "Descartes and Newton" (1728)

[T]he celebrated *Galileo* . . . was put in the inquisition for six
years, and put to the torture, for saying, that *the earth moved*.

   —Giuseppe Baretti, *The Italian Library* (1757)

[T]o say that Galileo was tortured is not a reckless claim, but it
is simply to repeat what the sentence says. To specify that he was
tortured about his intention is not a risky deduction, but it is,
again, to report what that text says. These are observation-
reports, not magical intuitions; proved facts, not cabalistic
introspections.

   —Italo Mereu, *History of Intolerance in Europe* (1979)

In the early years of the seventeenth century the Italian mathematician and natural philosopher Galileo Galilei (1564–1642)
openly advocated the theory of the earth's motion elaborated
in Nicolaus Copernicus's book *On the Revolutions of the Heavenly Spheres* (1543). As a result, he was persecuted, tried, and
condemned by the Catholic church. He spent the last nine years
of his life under house arrest in his villa outside Florence. But

was he imprisoned and tortured as the authors above, and countless others, have alleged?[1]

Galileo did not begin advocating Copernicanism until 1609. Before then, he was acquainted with Copernicus's work and appreciated the fact that it contained a novel and significant argument for the earth's motion. Galileo had been working on a new theory of motion and had intuited that the Copernican theory was more consistent with the new physics than was the geostatic theory. But he had not published or articulated this intuition. Moreover, he was acutely aware of the considerable evidence against Copernicanism stemming from direct sense experience, astronomical observation, traditional physics, and scriptural passages. Accordingly, he judged that the anti-Copernican arguments far outweighed the pro-Copernican ones.

In 1609, however, he perfected the newly invented telescope, and in the next few years by its means he made several startling discoveries: mountains on the moon, innumerable stars besides those visible with the naked eye, dense collections of stars in the Milky Way and nebulas, four satellites around Jupiter, the phases of Venus, and sunspots. He described them in *The Sidereal Messenger* (1610) and *Sunspots Letters* (1613).

As Galileo started showing that the new telescopic evidence rendered Copernicanism a serious contender for real physical truth, he came increasingly under attack from conservative philosophers and clergymen. They argued that he was a heretic because he believed in the earth's motion and the earth's motion contradicted Scripture. Galileo felt he could not remain silent and decided to refute the biblical arguments against Copernicanism. He wrote his criticism in the form of long, private letters, in December 1613 to his disciple Benedetto Castelli and in spring 1615 to the grand duchess dowager Christina.

Galileo's letter to Castelli provoked the conservatives even further, and so in February 1615 a Dominican friar filed a written complaint against Galileo with the Inquisition in Rome. The resulting investigation lasted about a year. Galileo himself was not summoned to Rome, partly because the key witnesses exonerated him, partly because his critical letters had not been published, and partly because his publications contained neither a categorical assertion of Copernicanism nor a denial of the scientific authority of Scripture.

In December 1615, however, Galileo went to Rome of his own accord to defend the Copernican theory. Despite winning the intellectual arguments, his practical effort failed. In February 1616, Cardinal Robert Bellarmine (in the name of the Inquisition) gave Galileo a private warning forbidding him to hold or defend the view that the earth moved. Galileo agreed to comply. In March, the Index of Prohibited Books (the department charged with book censorship) published a decree, without mentioning Galileo, that declared that the earth's motion was physically false and contradicted Scripture and that Copernicus's book was banned until revised.

Until 1623, when Cardinal Maffeo Barberini became Pope Urban VIII, Galileo kept quiet about the forbidden topic. Since Barberini was an old admirer, Galileo felt freer and decided to write a book that would defend Copernicanism indirectly and implicitly. Thus he wrote a dialogue featuring three characters engaged in a critical discussion of the cosmological, astronomical, physical, and philosophical aspects of Copernicanism but avoiding the biblical or theological ones. Published in 1632, this *Dialogue* showed that the arguments favoring the earth's motion were stronger than those favoring the geostatic view. Galileo apparently felt that the book did not "hold" the theory of the earth's motion because it was not claiming that the favorable arguments were conclusive; it was not "defending" this theory because it was a critical examination of the arguments on both sides.

Galileo's enemies nevertheless complained that the book defended the earth's motion and so violated Bellarmine's warning and the Index's decree. A new charge also emerged: that the book violated a special injunction issued personally to Galileo in 1616, prohibiting him from discussing the earth's motion in any way whatever. Such a document had just been discovered in the file of the earlier proceedings. Thus he was summoned to Rome for trial, which began in April 1633.

At his first hearing Galileo admitted receiving from Bellarmine the warning that the earth's motion could not be held or defended. But he denied receiving a special injunction not to discuss the topic in any way whatever. In his defense he introduced a certificate he had obtained from Bellarmine in 1616, which mentioned only the prohibition to hold or defend. Galileo also claimed that the *Dialogue* did not defend the earth's motion but rather showed that the favorable arguments were not conclusive and so did not violate Bellarmine's warning.

In light of Bellarmine's certificate and of various irregularities with the special injunction, the Inquisition's officials tried out-of-court plea-bargaining: they promised not to press the most serious charge (violation of the special injunction) if Galileo would plead guilty to a lesser charge (transgression of the warning not to defend Copernicanism). Galileo agreed, and so at subsequent hearings (on April 30 and May 10) he admitted that the book was written in such a way as to give readers the impression of defending the earth's motion. However, he denied that this had been his intention and attributed his error to conceit.

The trial ended on June 22, 1633, with a harsher sentence than Galileo had been led to expect. The verdict found him guilty of a category of heresy intermediate between the most and the least serious, called "vehement suspicion of heresy." The objectionable beliefs were the astronomical thesis that the earth moves and the methodological principle that the Bible is not a scientific authority. He was forced to recite a humiliating "abjuration" retracting these beliefs. But the *Dialogue* was banned.[2]

The lengthy sentencing document also recounted the proceedings since 1613, summarized the 1633 charges, and noted Galileo's defense and confession. In addition, it provided two other extremely important details. The first described an interrogation: "Because we did not think you had said the whole truth about your intention, we deemed it necessary to proceed against you by a rigorous examination. Here you answered in a Catholic manner, though without prejudice to the above-mentioned things confessed by you and deduced against you about your intention." The second imposed an additional penalty: "We condemn you to formal imprisonment in this Holy Office at our pleasure."[3]

The text of the Inquisition's sentence and Galileo's abjuration were the only trial documents publicized at the time. Indeed, the Inquisition sent copies to all provincial inquisitors and papal nuncios, requesting them to disseminate the information. Thus news of Galileo's fate circulated widely in books, newspapers, and one-page flyers. This unprecedented publicity resulted from the express orders of Pope Urban, who wanted Galileo's case to serve as a negative lesson to all Catholics and to strengthen his own image as an intransigent defender of the faith.[4]

The prison clause in the sentence clearly stipulated that Galileo was to be imprisoned in the jail at the Inquisition palace in Rome for an indefinite period, as long as the authorities wanted. Anyone reading or hearing of the sentence would naturally assume that the Inquisition had carried out the sentence it had imposed.

Although the sentence did not use the word *torture,* it did speak of a "rigorous examination," a technical term connoting torture. Moreover, the passage gave a reason why the judges had decided to subject Galileo to a rigorous examination: after the various interrogations, including his confession (to having defended Copernicanism), they had doubts about whether his transgression had been intentional (thus aggravating the crime) or inadvertent (as he claimed). In Inquisition practice (as well as in lay criminal courts) such doubts justified the administration of torture (in order to resolve them). The passage informed readers that Galileo had passed the rigorous examination when it stated that he "answered in a Catholic manner." That is, Galileo had answered like a good Catholic, who would not intentionally do something the church had forbidden. Finally, the passage clarified, again in accordance with inquisitorial practice, that Galileo's denial of a malicious intention (his "Catholic answers") did not undermine the other incriminating evidence coming from his confession and other sources (for example, opinions on the *Dialogue* written by three consultants). Readers of the sentence acquainted with legal terminology and practice understandably concluded that Galileo had suffered torture at the hands of his inquisitors.[5]

The impression that Galileo had been imprisoned and tortured remained plausible as long as the principal evidence available about Galileo's trial came from these documents, the sentence and abjuration. The story remained unchanged until—after about 150 years for the prison thesis and about 250 years for the torture thesis—relevant documents came to light showing that Galileo had suffered neither.

The new information about imprisonment comes from correspondence in 1633, primarily from the Tuscan ambassador to Rome (Francesco Niccolini) to the Tuscan secretary of state in Florence, and secondarily that to and from Galileo himself. The Tuscan officials were especially interested in Galileo because he was employed as the chief mathematician and philosopher to the grand duke of Tuscany, had dedicated the *Dialogue* to him, and had successfully sought his help in publishing the book in Florence. Thus the Tuscan government treated the trial like an affair of state, with Niccolini constantly discussing the situation directly with the pope at their regular meetings and sending reports to Florence. Moreover, Galileo was on very friendly terms with Niccolini and his wife.[6]

The 1633 correspondence, which surfaced in 1774–1775, shows that Galileo, answering the Inquisition's summons, left Florence on January 20 and arrived in Rome on February 13. The Inquisition allowed him to lodge at the Tuscan embassy (which served also as the Niccolinis' residence) on condition that he remain in seclusion until the proceedings started. On April 12 Galileo went to the Inquisition palace for his first interrogation. He stayed there for the next eighteen days while undergoing further interrogations, but he was put up in the prosecutor's six-room apartment, together with a servant, who brought him meals twice a day from the Tuscan embassy. On April 30, after his second deposition was recorded and signed, Galileo returned to the embassy, where he remained for fifty-one days, interrupted by a visit to the Inquisition palace on May 10 to give a third deposition. On Monday, June 20, he was summoned to appear in court the following day. On Tuesday he underwent the rigorous examination—and remained at the Inquisition palace until the evening of June 24. It is unclear whether he was held in a prison cell or permitted to use the prosecutor's apartment. On June 22 he appeared at the convent of Santa Maria sopra Minerva for sentencing and abjuration. Two days later Galileo moved from the Inquisition palace to Villa Medici in Rome, a sumptuous palace owned by the grand duke of Tuscany. On June 30 the pope granted Galileo permission to travel to Siena to live under house arrest at the residence of the archbishop, a good friend of Galileo's. The archbishop hosted him for five months. In December 1633 Galileo returned to his own villa in Arcetri, near Florence, where he remained—except for a brief period in 1638, when he resided within the city limits of Florence—under house arrest until his death in 1642.

With the possible exception of three days (June 21–24, 1633), Galileo was never held in prison, either during the trial (as was universal custom) or afterward (as the sentence decreed). Even for those three days he likely lodged in the prosecutor's apartment, not in a cell. The explanation for such unprecedentedly benign treatment is not completely clear but includes the following factors: the protection of the Medici, Galileo's celebrity status, and the love-hate attitude of Pope Urban, an erstwhile admirer.

The evidence for staying out of jail tells us nothing about Galileo's success in avoiding torture. The resolution of this question had to wait until the trial proceedings were published and assimilated in the late nineteenth century.[7] Two documents proved crucial.[8] The first was the minutes of the Inquisition meeting of June 16, 1633, chaired by the pope. After various reports and opinions were heard and after considerable discussion,

> His Holiness decided that the same Galileo is to be interrogated even with the threat of torture; and that if he holds up, after a vehement abjuration at a plenary meeting of the Holy Office, he is to be condemned to prison at the pleasure of the Sacred Congregation, and he is to be enjoined that in the future he must no longer treat in any way (in writing or orally) of the earth's motion or sun's stability, nor of the opposite, on pain of relapse; and that the book written by him and entitled *Dialogo di Galileo Galilei Linceo* is to be prohibited.

This preview of the actual sentence mentions a novel procedure: interrogation under the threat of torture. The minutes of the interrogation, dated June 21 and signed by Galileo, reveal that the commissary asked him several times whether he held the Copernican theory of the earth's motion; each time Galileo denied having done so after the condemnation of that doctrine in 1616. The exchange between Galileo and his inquisitors is worth quoting in full:

> *Q:* Having been told that from the book itself and the reasons advanced for the affirmative side, namely that the earth moves and the sun is motionless, he is presumed, as it was stated, that he holds Copernicus's opinion, or at least that he held it at the time, therefore he was told that unless he decided to proffer the truth, one would have recourse to the remedies of the law and to appropriate steps against him.
>
> *A:* I do not hold this opinion of Copernicus, and I have not held it after being ordered by injunction to abandon it. For the rest, here I am in your hands; do as you please.
>
> *Q:* And he was told to tell the truth, otherwise one would have recourse to torture.
>
> *A:* I am here to obey, but I have not held this opinion after the determination was made, as I said.
>
> And since nothing else could be done for the execution of the decision, after he signed he was sent to his place.
>
> I, Galileo Galilei, have testified as above.

This deposition leaves no doubt that Galileo was *threatened* with torture during the June 21 interrogation. But there is no evidence that he was *actually* tortured, or that his accusers planned actually to torture him. Apparently, the "rigorous examination" mentioned in the sentence meant interrogation with the threat of torture, not interrogation under actual torture.

The most common and relevant torture in Rome at the time was "torture of the rope." This consisted of tying the victim's wrists together behind the back, then tying the joined wrists to the end of a long rope that went around a pulley hung from the ceiling. The executioner held the other end of the rope in such a way that the victim could be hoisted in midair and left hanging for different periods (a standard rule specified a maximum of one hour). To increase the strain, weights of various amounts could be attached to the victim's feet. Alternatively, the victim would be dropped from various heights, just short of hitting the ground; the greater the height of the drop, the greater the pain in the victim's arms and joints (in fact, the numerical values of the distance dropped provided a quantitative measure of the severity of the torture).[9]

Because of the severity of the torture of the rope, we can be fairly certain that Galileo was not tortured this way. Given his advanced age of sixty-nine years and his frailty, he would have suffered permanent injury to his arms and shoulders, but there is no evidence of this. Moreover, if he had been tortured, it would have happened on June 21, leaving him in no condition to attend the sentencing and recite the abjuration on the twenty-second. Furthermore, Inquisition rules required that the torture session, including the victim's cries and groans, be recorded, but the proceedings contain no such minutes. Inquisition rules also stipulated that confessions obtained during torture be ratified twenty-four hours later, outside the torture chamber, but there is no record of ratification. And before a defendant could be tortured, there had to be a formal vote by the Inquisition consultants recommending it, as well as a decree to that effect by the inquisitors; but no minutes indicate that these steps were taken in Galileo's case.[10]

In addition, Inquisition authorities in Rome rarely practiced torture, further reducing the likelihood that Galileo experienced this punishment. Inquisitorial rules exempted old or sick people (along with children and pregnant women) from torture, and Galileo was not only elderly but suffering from arthritis and a hernia. The rules also spared clerics, and we now know that Galileo had received the clerical tonsure (a ceremonial haircut given to men being inducted into the clergy) on April 5, 1631, in order to benefit from an ecclesiastical pension. For reasons that may easily be guessed, the rules of torture stipulated that defendants could not be tortured unless a period of ten hours had elapsed since their last meal; but the known pace of the trial did not leave a gap of this length. Finally, another rule held that defendants could not be tortured during the investigation of an alleged crime unless the transgression was serious enough to require corporal punishment. Galileo's alleged crimes fell short of formal heresy, which would have justified corporal punishment; therefore, torturing him would have been inappropriate.[11]

Of course, all of the aforementioned rules and practices were subject to exception. For example, although old men could not be subjected to the rope torture, they could undergo the torture of fire to the feet. And although clerics could not be tortured by laymen, they could be tortured by other clerics. Moreover, the rules were often abused or disregarded by individual officials.[12] Additionally, several intermediate steps existed between the two extremes of threat during an interrogation outside the torture chamber and actual torture with the infliction of physical pain in the torture chamber—from showing the defendant the instruments of torture, to undressing the victim and then tying him to the instruments in preparation, and so on. The term *territio realis* (meaning "real intimidation," as distinct from *territio verbalis,* or "verbal threat") was used to refer to these intermediate steps. Some scholars have speculated that Galileo was subjected to *territio realis.* This version of the torture thesis is not incompatible with the June 16 papal orders or with the fact that Galileo showed no signs of shoulder dislocation after June 21, that he had enough physical strength to attend the sentencing and abjuration on June 22, that there was no ratification of the confession under actual torture, and that there was no consultants' vote or Inquisitors' decree for torture. However, it is inconsistent with the June 21 deposition, which contains no description of those intermediate steps. Hence this version of the torture thesis presupposes the inauthenticity of that deposition.[13]

One could object that even if Galileo was not tortured physically, the treatment he received amounted to moral (or psychological) torture, that is, the threat of torture in the last interrogation and the perpetual house arrest after the condemnation. Indeed, since the middle of the nineteenth century many authors have maintained the moral-torture thesis.[14] But the argument for moral torture starts down a semantic slippery slope with no end in sight.

In view of the available evidence, the most tenable position is that Galileo underwent an interrogation with the threat of torture but did not undergo actual torture or even *territio realis.* Although he remained under house arrest during the 1633 trial and for the subsequent nine years of his life, he never went to prison. We should keep in mind, however, that for 150 years after the trial the publicly available evidence indicated that Galileo had been imprisoned, and for 250 years the evidence indicated that he had been tortured. The myths of Galileo's torture and imprisonment are thus genuine myths: ideas that are in fact false but once seemed true—and continue to be accepted as true by poorly educated persons and careless scholars.

MYTH 9

THAT CHRISTIANITY GAVE BIRTH
TO MODERN SCIENCE

Noah J. Efron

Faith in the possibility of science, generated antecedently to the
development of modern scientific theory, is an unconscious
derivative from medieval theology.

> —Alfred North Whitehead, *Science and the Modern
> World* (1925)

[T]he fundamental paradigm of science: its invariable stillbirths
in all ancient cultures and its only viable birth in a Europe which
Christian faith in the Creator had helped to form.

> —Stanley L. Jaki, *The Road of Science and the Ways
> to God* (1978)

[T]heological assumptions unique to Christianity explain why
science was born only in Christian Europe. Contrary to the
received wisdom, religion and science not only were compatible;
they were inseparable . . . *Christian theology was essential for
the rise of science.*

> —Rodney Stark, *For the Glory of God* (2003)

A Newtonian might put it this way: for every myth there is an
equal and opposite myth. Consider popular accounts of Christianity's relationship to science. Everyone is familiar with the
myth that popes, bishops, priests, ministers, and pastors all saw
it as a sacred duty to silence scientists, stymie their inquiries, and

stifle their innovations. Lately, a new account of Christianity's link to science has been put forth, opposite in attitude to the first but equally bold and, in the end, equally wrong. In this account, not only did Christianity not quash science, but it and it alone gave birth to modern science and nurtured it to maturity. And the world is a far better place for it. As the Baylor University sociologist Rodney Stark has recently claimed,

> Christianity created Western Civilization. Had the followers of Jesus remained an obscure Jewish sect, most of you would not have learned to read and the rest of you would be reading from hand-copied scrolls. Without a theology committed to reason, progress, and moral equality, today the entire world would be about where non-European societies were in, say, 1800: A world with many astrologers and alchemists but no scientists. A world of despots, lacking universities, banks, factories, eyeglasses, chimneys, and pianos. A world where most infants do not live to the age of five and many women die in childbirth—a world truly living in "dark ages."[1]

As Stark sees it, chimneys and pianos, and all the more so chemistry and physics, owe their existence to Catholics and Protestants.

To be fair, the claim that Christianity led to modern science captures something true and important. Generations of historians and sociologists have discovered many ways in which Christians, Christian beliefs, and Christian institutions played crucial roles in fashioning the tenets, methods, and institutions of what in time became modern science.[2] They found that some forms of Christianity provided the *motivation* to study nature systematically; sociologist Robert Merton, for example, argued seventy years ago that Puritan belief and practice spurred seventeenth-century Englishmen to embrace science.[3] Scholars still debate what Merton got right and what he got wrong, and in the intervening years they have drawn a far more detailed portrait of the varied nature of the religious impetus to study nature. Although they disagree about nuances, today almost all historians agree that Christianity (Catholicism as well as Protestantism) moved many early-modern intellectuals to study nature systematically.[4] Historians have also found that notions borrowed from Christian *belief* found their ways into scientific discourse, with glorious results; the very notion that nature is lawful, some scholars argue, was borrowed from Christian theology.[5] Christian convictions also affected how nature was studied. For example, in the sixteenth and seventeenth centuries, Augustine's notion of original sin (which held that Adam's Fall left humans implacably damaged) was embraced by advocates of "experimental natural philosophy." As they saw it, fallen humans lacked the grace to understand the workings of the world through cogitation alone, requiring in their disgraced state painstaking experiment and observation to arrive at knowledge of how nature works (though our knowledge even then could never be certain). In this way, Christian doctrine lent urgency to experiment.[6]

Historians have also found that changing Christian approaches to interpreting the Bible affected the way nature was studied in crucial ways. For example, Reformation leaders disparaged allegorical readings of Scripture, counseling their congregations to read Holy Writ literally. This approach to the Bible led some scholars to change the way they studied nature, no longer seeking the allegorical meaning of plants and animals and instead seeking what they took to be a more straightforward description of the material world.[7] Also, many of those today considered "forefathers" of modern science found in Christianity legitimation of their pursuits. René Descartes (1596–1650) boasted of his physics that "my new philosophy is in much better agreement with all the truths of faith than that of Aristotle."[8] Isaac Newton (1642–1727) believed that his system restored the original divine wisdom God had provided to Moses and had no doubt that his Christianity bolstered his physics—and that his physics bolstered his Christianity.[9] Finally, historians have observed that Christian churches were for a crucial millennium leading patrons of natural philosophy and science, in that they supported theorizing, experimentation, observation, exploration, documentation, and publication.[10]

They did this in some circumstances directly, in church institutions such as the renowned Jesuit seminary, the *Collegio Romano,* and in other circumstances less directly, through universities supported in part or full by the church.

For all these reasons, one cannot recount the history of modern science without acknowledging the crucial importance of Christianity. But this does not mean that Christianity and Christianity alone produced modern science, any more than observing that the history of modern art cannot be retold without acknowledging Picasso means that Picasso created modern art. There is simply more to the story than that.

For one thing, Christian ideas about nature were not exclusively *Christian* ideas. Especially in the first centuries of Christian history, the views and sensibilities of Christians were shaped by the "classical tradition," an intellectual heritage that included art, rhetoric, history, poetry, mathematics, and philosophy, including the philosophy of nature. This tradition could at times appear threadbare; many of the original Greek texts were lost, and only a portion of those that remained were translated into Latin and thereby accessible to Christian scholars (who, in increasing numbers over the generations, were never taught Greek). Also, the classical tradition, which was after all pagan in origin, was a matter about which church fathers were understandably ambivalent. Still, the imprint of Greek and Roman ideas on Christian intellectuals remained vivid; they provided the starting point for nearly all inquiries into nature until the start of the modern era. For many centuries, Aristotle's philosophy was knit most tightly into the woof and warp of Christian theology. In time, and especially during the Renaissance, Platonic and Neoplatonic philosophies came fitfully to challenge Aristotelianism in varied realms. Just which classical philosophers influenced which Christian intellectuals in which ways is a story of exquisite intricacy, too varied to recount in this short chapter.[11] But the general point is clear. Excluding the place of classical philosophers from an account of the history of modern science is an act of intellectual appropriation of breathtaking arrogance, and one that the forefathers of modern science themselves would never have agreed to. In the sixteenth century, Nicolaus Copernicus's (1473–1543) view that the sun is at the center of the universe was often called the "Pythagorean hypothesis," and Galileo Galilei (1564–1642) and Johannes Kepler (1571–1630) both traced the roots of their innovations back to Plato. These men and their contemporaries all knew what some today have forgotten, that Christian astronomers (and other students of nature) owe a great debt to their Greek forebears.

This was not the only debt outstanding for Christian philosophers of nature. They had also benefited directly and indirectly from Muslim and, to a lesser degree, Jewish philosophers of nature who used Arabic to describe their investigations. It was in Muslim lands that natural philosophy received the most careful and creative attention from the seventh to the twelfth century.[12] The reasons for this had much to do with the rapid spread of Islamic civilization over vast territories in which other cultures had long before laid down deep roots. By virtue of its geography alone, Islam became "the meeting point for Greek, Egyptian, Indian and Persian traditions of thought, as well as the technology of China."[13] This was an asset of incalculable value. For one thing, practical know-how (like how to produce paper) spread from culture to culture. For another, the multiplicity of intellectual and cultural traditions absorbed by Islam were synthesized in startling and creative ways, lending Islamic culture a richness and authority far beyond what one might expect to find in a relatively young civilization. Indeed, by the start of the ninth century, great numbers of Greek, Indian, and Persian books of philosophy and natural philosophy had been translated into Arabic, and by the year 1000 the library of ancient writings available in Arabic was vastly superior to the works available in Latin or any other language. It included a great deal of Indian astronomy and mathematics (translated from Sanskrit and Pahlavi), most of the Hellenistic corpus, and much Greek philosophy. These translations were of immense worth for philosophers of nature of later generations, but the real importance of Arabic-language scholarship went far beyond translation alone. Muslim scholars added sophisticated commentaries and glosses to Greek texts and wrote original essays that advanced every major field of inquiry, mathematics, astronomy, optics, and above all medicine. They developed intricate instruments of observation, built (with the support of caliphs) massive observatories, and collected volumes of observations that retained their value for astronomers for long centuries.

Many of these Muslim achievements were, in time, eagerly adopted by Christian philosophers of nature. As Christians slowly reconquered much of Spain and Sicily from Muslim rulers in the twelfth and thirteenth centuries, they came into more intimate contact with the large corpus of Arabic texts, translations, and original treatises and discovered as well Greek texts that had previously been lost to them. Christian scholars, aided on occasion by Jews, translated many of these texts into Latin, and this grand body of new materials forever changed the course of Christian philosophy of nature.[14] Recently, historians have made the case that the direct influence of Islamic natural philosophy on Christians continued unabated up to the start of the modern period. Pointing to the fact that renowned Christian scholars such as the Parisian Guillaume Postel (1510–1581) read and annotated advanced Arabic astronomic texts, they hypothesize that Copernicus himself may have borrowed his revolutionary astronomy from a famous Damascus astronomer named Ibn al-Shātir (ca. 1305–1375), who had proposed a similar system generations earlier. "With Poland, where Copernicus was born, being so close to the borders of the Ottoman empire at the time, and with the free flow of books, trade and scholars across the Mediterranean through the northern Italian cities, where Copernicus received his education, we must suspect that there were many people like Postel who could have advised or even tutored Copernicus on the contents of Arabic astronomical texts."[15] Historians may quarrel about whether Copernicus cribbed his sun-centered system from an obscure Muslim treatise, but they agree that Islam's impact on Christian philosophy of nature was enduring.

Anthropologist Clifford Geertz once told a story about an Englishman in India who, "having been told the world rested on a platform which rested on the back of an elephant which rested in turn on the back of a turtle, asked . . . what did the turtle rest on? Another turtle. And that turtle? 'Ah, Sahib, after that it is turtles all the way down.' "[16] Science is a bit that way. Modern science rests (somewhat, anyway) on early-modern, Renaissance, and medieval philosophies of nature, and these rested (somewhat, anyway) on Arabic natural philosophy, which rested (somewhat, anyway) on Greek, Egyptian, Indian, Persian, and Chinese texts, and these rested, in turn, on the wisdom generated by other, still earlier cultures. One historian has called this twisting braid of lineage "the dialogue of civilizations in the birth of modern science."[17] Recognizing that modern science grew out of the give-and-take among many cultures over centuries does not denigrate the crucial role of early-modern Protestants and Catholics in casting the molds in which modern science grew. Ignoring this fact obscures something of fundamental importance about modern science, however: the rich diversity of the cultural and intellectual soil deep into which its roots extend.

Even if you look no farther than Europe during "the Scientific Revolution" for the origins of modern science, religion is only part of what you will find. One historian has recently argued, for instance, that commerce had as much to do with the rise of modern science as Christianity did; "the values inherent in the world of commerce were explicitly and self-consciously recognized to be at the root of the new science by contemporaries." It was the carrot-and-stick of competitive trade that led "to countless efforts to find out matters of fact about natural things and to ascertaining whether the information was accurate and commensurable."[18] Further, the early-modern voyages of discovery and the rapid establishment of new maritime trade routes in short order flooded

Europe with new information, new goods, and even new plants and animals, all of which sparked new lines of inquiry and new theories about nature and, in particular, natural history. Thus, although we tend to think that the growth of science led to advances in technologies that created wealth and prosperity, the reverse was also true. Increased trade created a need for new technologies and for verified facts about nature that modern science came to provide.

Historians have also concluded that a great many other forces affected the growth of modern science in Europe. Some have found that the invention or importation of important technologies, like clocks and especially the printing press, gave a boost to the sorts of inquiry that in time developed into modern science. Others have found that changes in European political organization spurred the development of science in complicated ways, and still others have found that, as they developed, Europe's great legal systems influenced the development of both scientific theory and practice.[19] Early-modern Europe also saw the emergence of other secular institutions that came to play an important role in the growth of modern science. Scientific societies, for example, were established across the continent beginning in the seventeenth century. The founders and early members were pious Christians of one or another denomination, but they wished the fellowships they built to transcend religious affiliation. Bishop Thomas Sprat (1635–1713), for instance, wrote of the founders of the Royal Society that they "have freely admitted Men of different Religions, Countries, and Professions of Life. This they were oblig'd to do, or else they would come far short of the Largeness of their own Declarations. For they openly profess, not to lay the Foundation of an *English, Scotch, Irish, Popish or Protestant* Philosophy; but a Philosophy of *Mankind.*"[20] Admittedly, Sprat imagined a society that companionably joined all sorts of *Christians;* but the ideal he articulated was broader than his imagination. Over time, this ideal weakened the links between science and Christian belief and practice.

By the eighteenth century, at least some of Europe's leading natural philosophers and natural historians did not see themselves as Christians at all. The Swiss mathematician Johann Bernoulli (1667–1748) and the French natural historian Georges-Louis Leclerc de Buffon (1708–1788), for instance, argued against Newton's physics for taking for granted a God who supplied the forces that draw bodies together. Having come to doubt the Christian God, they were convinced that nature can and must be described without reference to this God. Such men, inspired by what scholars have called the "radical enlightenment," remained a very small minority among those debating matters of physics, chemistry, biology, and the like. But their research and their views were undeniably part of the history of modern science.[21]

With time, as modern science became more firmly established, the cultural diversity of modern science grew stronger still. While Christianity continued into the nineteenth and twentieth centuries to motivate many scientists and to influence their ideas and behaviors, as time passed the impact of Christianity grew less and less public, and less inevitable. By the twentieth century, a large percentage of active scientists were not Christian at all; they were Jews, Hindus, Buddhists, Taoists, and, with growing frequency, avowed agnostics and atheists. With the passage of time, the *ethos* of science came to stand at odds with the particularist claims of any religious or ethnic group. By 1938, Robert Merton could declare as a simple matter of fact that "it is a basic assumption of modern science that scientific propositions 'are invariant with respect to the individual' and groups . . . Science must not suffer itself to become the handmaiden of theology or economy or state."[22]

When boosters insist that "Christianity is not only compatible with science, it created it," they are saying something about science, they are saying something about Christians, and they are saying something about everyone else.[23] About science, they are saying that it comes in only one variety, with a single history, and that centuries of inquiries into nature in China, India,

Africa, the ancient Mediterranean, and so on have no part in
that history. About Christians, they are saying that they alone
had the intellectual resources—rationality, belief that nature is
lawful, confidence in progress, and more—needed to make sense
of nature in a systematic and productive way. About everyone
else, they are saying that, however admirable their achievements
in other realms may be, they lacked these same intellectual resources.

Often enough, what these boosters really mean to say, sometimes straight out and sometimes by implication, is that Christianity has given the world greater gifts than any other religion.
Frequently, they mean to demonstrate simply that Christianity is
a better religion.

> Real science arose only once: in Europe. China, Islam, India, and an-
> cient Greece and Rome each had a highly developed alchemy. But
> only in Europe did alchemy develop into chemistry. By the same to-
> ken, many societies developed elaborate systems of astrology, but
> only in Europe did astrology lead to astronomy. Why? Again, the
> answer has to do with images of God.[24]

In this way of seeing things, only Christian images of God were
rich enough, optimistic enough, and rational enough to lead to
"real science." If we use modern science as a measuring stick,
Taoists, Buddhists, Hindus, Muslims, and pagans were all just
wannabes—picturesque perhaps, but lacking the right stuff.

This anything-your-religion-does-mine-can-do-better attitude
jiggers one part condescension with two parts self-congratulation,
and one wonders why some find it appealing. Yes, Christian belief, practice, and institutions left indelible marks on the history
of modern science, but so too did many other factors, including
other intellectual traditions and the magnificent wealth of natural
knowledge they produced. Assigning credit for science need not
be a zero-sum game. It does not diminish Christianity to recognize
that non-Christians, too, have a proud place in the history of
science.

It is also worth noting that science itself is an ambivalent legacy. In 1967, historian Lynn White, Jr., wrote in a famous essay, "The Historical Roots of Our Ecological Crisis," that "Somewhat over a century ago science and technology—hitherto quite separate activities—joined to give mankind powers which, to judge by many of the ecological effects, are out of control. If so, Christianity bears a huge burden of guilt."[25] White's thesis has been debated endlessly and with fervor for forty years, and scholars now agree (for all the reasons I described above) that whatever damage modern science and technology have caused cannot blithely be chalked up to Christianity. Debates of this sort lead us to a greater point, however. As bombs of awesome sophistication are deployed daily on distant battlefields, as the earth heats up and oceans rise, as bacteria gain resistance to antibiotics, there is value in seeing science for what it really is: a marvelous human creation of exquisite complexity and ingenuity, the effects of which have been both good and bad. As we enlist science to fashion lasting solutions for global problems (some of which were themselves caused by science itself), it is a consoling fact that science is neither the project nor the province of any single group to the exclusion of all others. For better and for worse, science is a *human* endeavor, and it always has been.

THAT THE SCIENTIFIC REVOLUTION LIBERATED
SCIENCE FROM RELIGION

Margaret J. Osler

At least one . . . dimension of the Scientific Revolution demands notice—a new relation between science and Christianity . . . From the point of view of science, it does not seem excessive to speak of its liberation. Centuries before, as European civilization had taken form out of the chaos of the dark ages, Christianity had fostered, molded, and hence dominated every cultural and intellectual activity. By the end of the seventeenth century, science had asserted its autonomy.

> —Richard S. Westfall, "The Scientific Revolution Reasserted" (2000)

It was unquestionably the rise of powerful new philosophical systems, rooted in the scientific advances of the early seventeenth century and especially the mechanistic views of Galileo, which chiefly generated that vast *Kulturkampf* between traditional, theologically sanctioned ideas about Man, God, and the universe and secular, mechanistic conceptions which stood independently of any theological sanction.

> —Jonathan I. Israel, *Radical Enlightenment: Philosophy and the Making of Modernity, 1650–1750* (2001)

The Scientific Revolution liberated science from religion. The new science separated spirit from matter. Reason and experiment replaced revelation as the source of knowledge of the world. After the Scientific Revolution, it was inevitable that God would eventually be pushed entirely out of nature and that science would deny the existence of God. These unsubstantiated claims have worked their way into the popular history of science and are frequently repeated. Journalists reporting on debates about evolution and creation, environmentalists searching for the sources of global warming, feminists writing critiques of science, and would-be prophets of New Age spirituality lamenting the disenchantment of the modern world—all repeat this mantra, reinforcing the belief that the seventeenth century witnessed a divorce of science from religion. Unfortunately for these commentators, a closer look at the history of that turbulent century reveals an entirely different story.

Science and *religion* did not have the same meanings then that they do today. The difference is particularly blatant with regard to science. There was no such creature as a scientist—the word did not even exist until the nineteenth century. The pursuit of knowledge about the world was called "natural philosophy." What was understood by this term? Seventeenth-century thinkers inherited this discipline from their medieval predecessors. Its scope derived from Aristotle's classification of the sciences, which had become entrenched in the curriculum of the medieval universities. Physics, or natural philosophy, was one of the theoretical sciences and dealt with things that are inseparable from matter but not immovable. In the medieval universities natural philosophy was part of the undergraduate curriculum, and its subject matter was treated without specific reference to church doctrine. Theology, however, was taught in a separate, graduate faculty. The study of natural philosophy included the study of the first causes of nature, change and motion in general, the motions of celestial bodies, the motions and transformations of the elements, generation and corruption, the phenomena of the upper atmosphere right below the lunar sphere, and the study of animals and plants. These subjects included consideration of God's creation of the world, the evidence of divine design in the world, and the immortality of the human soul. Despite the fact that natural philosophy and theology occupied separate places in the medieval curriculum, medieval natural philosophy was conditioned by theological presuppositions, and its conclusions pertained to important theological issues. Discussions of the causes of things, for example, included questions about the cause of the world and revolved around the divine creation of the world. Discussions of matter and change had implications for the interpretation of the Eucharist (particularly the claim that the bread and wine actually became the body and blood of Christ). Discussion of the nature of animals and how they differ from humans had direct bearing on questions about the immortality of the human soul.

Despite their widespread rejection of Aristotelianism, early-modern natural philosophers continued to deal with the same range of topics as their medieval predecessors. Natural philosophy encompassed many topics now considered theological or metaphysical—such as the immortality of the soul and the study of divine providence in nature—and excluded others—such as optics and astronomy, then known as "mixed mathematics"—that are now considered to be scientific disciplines. The word *science* retained its Scholastic meaning: *scientia* referred to demonstrative knowledge of the real essences of things. Although there were individual sciences, there was no general term *science* to refer to a whole category of knowledge.

During the seventeenth century, the growth of empirical investigation, especially in natural history, yielded less-than-certain knowledge about the world, making it increasingly difficult to assimilate natural philosophy into the Aristotelian demonstrative model for science. In addition, a skeptical crisis brought on by the almost simultaneous recovery of the writings of the ancient Greek skeptics and post-Reformation debates about authority in religion challenged the very idea of certainty as a possible standard for knowledge about the world. In light of these developments, many natural philosophers developed a theory of knowledge that emphasized observation, and they regarded their conclusions as merely probable, further distinguishing their endeavors from the Aristotelian model of *scientia*. As late as 1690, the philosopher John Locke (1632–1704) could write "that natural philosophy is not capable of being made a science."[1]

A different set of distinctions pertains to the terms *religion* and *theology*. *Religion* refers to doctrine, faith, and practice, whether or not these concepts are interpreted within an institutional setting. *Theology* refers to the enterprise of explaining the meaning of religious doctrines or practices, usually by employing philosophical concepts and arguments. For example, the Roman Catholic celebration of the Eucharist is a religious practice; the explanation of the real presence of Christ in the elements of the mass by the Thomist theory of transubstantiation is theological. Many developments discussed in the context of science and religion in the early-modern period are actually better described as issues concerning natural philosophy and theology.

Despite the facts that some thinkers sought to differentiate between the pursuits of natural philosophy and theology and that the Royal Society excluded discussions of religion and politics from its meetings, the close relationship between natural philosophy and theology is evident in almost every area of inquiry about the natural world during the Scientific Revolution.[2] The debates about the new heliocentric astronomy, the arguments for a new philosophy of nature to replace medieval Aristotelianism, the development of a new concept of the laws of nature, and discussions of the scope and limits of human knowledge were all infused with religious commitments and theological presuppositions.

The debates surrounding Copernican astronomy often reflected theological positions and hinged on the relative weights given to theological, philosophical, or astronomical claims. For example, the Lutheran intellectuals around Wittenberg tended to use Copernican methods for making astronomical calculations without accepting Copernicus's heliocentric cosmology. In many cases they were willing to accept a hypothetical interpretation of astronomical theory while insisting on a literal interpretation of Scripture. Lutheran theology, especially Lutheran ideas about providence, may have been one of Johannes Kepler's primary motives for insisting that God created a cosmos that exhibits geometrical order and arithmetic harmony.

Many natural philosophers, rejecting Aristotelianism, adopted some version of the mechanical philosophy. The mechanical philosophy aimed to explain all natural phenomena in terms of matter and motion. Mechanisms composed of microscopic particles of matter were thought to produce sensations of all the qualities we perceive in the physical world. The only causes acting in the mechanical world are the motions of these particles, which act by contact and impact. Although its critics feared that the mechanical philosophy would lead to materialism, the mechanical philosophers—such as Pierre Gassendi (1592–1655), René Descartes (1596–1650), and Robert Boyle (1627–1691), who were all devout Christians—limited the scope of their mechanization of nature by insisting on the existence of God, angels, demons, and an immortal human soul—all of which were spiritual, nonmaterial entities. The particular theories of matter that the mechanical philosophers adopted and their various ideas about the nature and scope of knowledge about the world reflected their theological presuppositions.

The mechanical philosophers adopted a new theory of causality along with the new theory of matter. They rejected the Aristotelian view that every change requires a complete explanation involving four causes: the formal cause (form), the material cause (matter), the efficient cause (the motions that bring about the change), and the final cause (the goal or purpose of the change). A simple example that illustrates the nature of the four Aristotelian causes is the construction of a house. The formal cause is the architect's plan, the blueprint. The material cause is the matter out of which the house is built—the two-by-fours, pipes, wires, roofing material, dry wall, and so on. The efficient cause is the activity of the workmen who actually construct the house. And the final cause or purpose of the house is to provide shelter. In the natural world—in the absence of an intelligent agent—changes are also explained in terms of the four causes. Consider the growth of an oak tree. The formal cause is the actualization of the form of the oak that exists potentially in the acorn and that is actualized in the development of the oak tree. The material cause consists of the water and earth and other matter of which the tree consists. The efficient cause is the actualization of the form from potential to actual oak tree. And the final cause is the production of an offspring that resembles its parents—that is to say, the actualization of the form of oak. In the case of biological examples, the formal, final, and efficient causes are often the same.

The mechanical philosophers reduced all causality to efficient causes. Nevertheless, close examination of early-modern texts has refuted the claim that the adoption of the mechanical philosophy automatically involved rejection of teleological (or goal-directed) explanations and, more generally, paved the way to materialism, deism, and atheism. Virtually all the mechanical philosophers claimed that God had created matter and had set it into motion. God infused his purposes into the creation either by programming the motions of the particles or by creating particles with very particular properties. Consequently, even a mechanical world had room for purpose and design. For example, Pierre Gassendi, who tried to make Greek atomism compatible with Christian theology, asserted there is in fact a role for final causes in physics—contrary to Francis Bacon (1561–1626) and René Descartes, both of whom had ruled them out; and Robert Boyle published an entire treatise on the role of final causes in natural philosophy. Isaac Newton (1642–1727) explicitly endorsed the appeal to final causes and argued that natural philosophy, properly pursued, leads to knowledge of the Creator. All of these natural philosophers reinterpreted the term *final cause* to refer to God's purposes imposed on the creation rather than to innate, goal-directed actions.

Another theme common to early-modern discussions about the possibility of human knowledge of the creation was that expressed by the metaphor of God's two books: the book of God's word (the Bible) and the book of God's work (the created world). Natural philosophers regarded both books as legitimate sources of knowledge. Early in the seventeenth century, Galileo Galilei (1564–1642) appealed to this metaphor in the context of a discussion of the relative importance of studying the Bible and observing natural phenomena: "the holy Bible and the phenomena of nature proceed alike from the divine Word, the former as the dictate of the Holy Ghost and the latter as the observant executrix of God's commands."[3]

Despite the claims of some modern commentators that the seventeenth century witnessed a sea change in the relationship between science and religion—especially after Newton's theory of universal gravitation seemed to reduce the cosmos to a mathematical equation (granted, a very powerful one)—the same metaphor served pretty much the same purpose at the end of the century. Newton himself took seriously both God's work and God's word, as he demonstrated by devoting even more effort to the understanding of Scripture than he did to the natural world. Boyle, too, argued that the study of Scripture as well as nature reveals truths about religion and the creation, respectively.[4]

Most importantly, the entire enterprise of studying the natural world was embedded in a theological framework that emphasized divine creation, design, and providence. These themes are prominent in the writings of almost all the major seventeenth-century natural philosophers. Boyle, Newton, and the naturalist John Ray (1627–1705) believed that the study of the created world provides knowledge of the wisdom and intelligence of the Creator, and they used the argument from design to establish God's providential relationship to his creation. Newton, whose physics historians have traditionally regarded as the culmination of the Scientific Revolution, shared these concerns. He clearly believed that theology is an intrinsic part of natural philosophy.

"For all discourse about God is derived through a certain similitude from things human, which while not perfect is nevertheless a similitude of some kind . . . [A]nd to treat of God from phenomena is certainly a part of 'natural' philosophy."[5]

Study of the created world produced knowledge of the phenomena and the laws of nature and revealed God's relationship to his creation. Natural philosophers disagreed about exactly how God relates to the world. Some, like Descartes, believed that once God created the laws of nature he could no longer change them. Others, like Gassendi and Boyle, insisted that God's power is not limited by anything he creates and so the laws of nature are nothing more than descriptions of the regular course of events, into which God is free to intervene at will. These different theological positions had implications for the methods adopted by natural philosophers. Those, like Descartes, who believed that God created necessary laws of nature also believed that knowledge of at least some aspects of the world could be known by purely rational means, that is, without empirical test. Others, like Gassendi, who believed that everything that God creates remains contingent on his free will, thought that the only way to discover facts about the world is by observation, because God could change the course of events at any moment. Such considerations continued to absorb the attention of natural philosophers throughout the seventeenth century, culminating in an epistolary debate between Newton's spokesman Samuel Clarke (1675–1729) and his rival in mathematics, Gottfried Wilhelm Leibniz (1646–1716). Much of this debate centered on the question of how God designed the world and the extent to which he intervenes in its workings.

Many modern commentators, not well informed about the realities of early-modern natural philosophy, assume that these discussions of God, the divine attributes, and the stated interconnections between natural philosophy and theology were simply lip service paid to the coercive religious authorities of the day. They assume that Galileo's conflict with the Roman Catholic Church was paradigmatic of the relationship between science and religion at the time and that scientists were struggling to break free from the stranglehold of ecclesiastical authority. But even the Galileo case does not fit into this mold. Galileo's troubles stemmed from his views on the relative authority of Scripture and science and on disagreement with principles of biblical interpretation, not from the Catholic church's blanket opposition to science. In the context of the church's strong reaction against the Reformation, Galileo's seemingly reasonable position was fraught with unwittingly controversial claims. Moreover, the Catholic church was one of the great patrons of the sciences in the seventeenth century, and many members of the Society of Jesus made significant contributions to astronomy and the natural philosophy of the day. (See Myths 2, 8, and 11.) Furthermore, in post-Reformation Europe, there was no longer a single church to dictate party line to scholars.

Seventeenth-century natural philosophers were not modern scientists. Their exploration of the natural world was not cut off from their religious views and theological assumptions. That separation came later. Reading the past from the standpoint of later developments has led to serious misunderstandings of the Scientific Revolution. For many of the natural philosophers of the seventeenth century, science and religion—or, better, natural philosophy and theology—were inseparable, part and parcel of the endeavor to understand our world.

THAT CATHOLICS DID NOT CONTRIBUTE

TO THE SCIENTIFIC REVOLUTION

Lawrence M. Principe

Roman Christianity and Science are recognized by their
respective adherents as being absolutely incompatible; they
cannot exist together . . . For Catholicism to reconcile itself to
Science . . . there is a bitter, mortal animosity to be overcome.

   —John William Draper, *History of the Conflict between
   Religion and Science* (1874)

There can simply be no doubt that Protestant and bourgeois
milieux have encouraged talent and ambition to rise through
science, and that Catholic and aristocratic milieux have inhibited
the development of scientists.

   —Charles C. Gillispie, *The Edge of Objectivity* (1960)

The twin assertions that the Catholic church opposed science
and that Catholics themselves contributed little to its development are widespread. These assertions are often casually assumed in a wide range of publications and among many parts of
the general public. Yet there is very little truth in either statement. Neither claim emerges from historical study or historical
sources. Instead, both are in large measure the products of self-
serving political or nationalistic rhetoric and old-fashioned
xenophobia. Let's look first at the creation and perpetuation of
these myths and then examine the contrasting testimony of the
historical record.

Perhaps the most influential and most frequently published book ever written on the subject of science and religion, John William Draper's *History of the Conflict between Religion and Science* (1874), is little more than a thinly disguised anti-Catholic rant. So slanted and hysterical that it is difficult for an educated person today to read it without smirking, Draper's work plays fast and loose with the facts and contains barely a shred of reliable historical information. Nonetheless, it continues to be read and cited by some, and a generation ago even some historians of science referenced it uncritically. More significantly, its claims have become "common knowledge" for a great number of people. Curiously enough, Draper's notions function mostly to perpetuate political phobias and discredited ideas of nineteenth-century America. Part of his anti-Catholicism is typical of that widespread sentiment in anglophone cultures, intensified in America by the anti-immigrant fears of nineteenth-century American Protestants, wary of the influx of Catholic immigrants, and part may well lie in personal animosity toward his sister, a convert to Catholicism.[1]

Yet anti-Catholicism is not a purely nineteenth-century phenomenon; it has famously been called "the most deeply held bias in the history of the American people."[2] Indeed, it continues to exist today in forms and places where racial prejudice and anti-Semitism would never be tolerated. This deep-seated attitude continues to reinforce and perpetuate the old myths about Catholicism and science.

America's anti-Catholicism was inherited from Britain, where it has been well established since the seventeenth century. Already in the 1640s, Galileo's misfortunes were being used in England to support "anti-papist" sentiments.[3] But even on the Continent, and in majority-Catholic countries, a related phenomenon developed in the eighteenth and nineteenth centuries. As various political and social movements came to oppose the secular power of the Catholic church, anticlericalism flourished. Since these movements tended to march under a banner of progressivism, they were supportive of the era's great emblem of progress: science and technology. Hence, it was easy—and politically expedient—to create the impression that the Catholic church was, and always had been, opposed to science, technology, and progress. In nineteenth-century Italy, the mythologizing of Giordano Bruno and Galileo into nationalist and anticlerical heroes (mentioned in Myth 7) was not unrelated to the political goals of the Risorgimento (the unification of Italy), which required the dismantling of papal temporal power.

Additionally, the very act of living in an English-speaking culture has served to eclipse or distort the role of Catholics—and to some extent of all Continentals in general—in the history of science. This result is due to the Anglocentric nature of so many accounts of the history of science written by English speakers. Once-prominent viewpoints, such as the "Merton thesis" (which claimed that the rise of science was linked to the rise of Puritanism), emerged from this unwarrantable Anglocentrism and reinforced it. Thus part of the task for the history of science today lies in transcending the boundaries of language and nationalism to provide a more balanced and accurate picture of the pan-European and transconfessional origins of modern science.

It would of course be absurd to claim that there have been no instances of Catholic laymen or clerics opposing scientific work in some form or other. Without question, such examples can be found, and quite easily. Yet it would be equally absurd to extend these examples of opposition—no matter how ignorant or ill-conceived—to the Catholic church or to Catholics as a whole. This act would be to commit the historical sin of overgeneralization, that is, the unwarrantable extension of the actions or thought of one member of a collective body to the entire body as a whole. (For example, there are apparently American flat-earthers alive today, yet it is not correct then to say that twenty-first-century Americans in general believe that the earth is flat.)

The Catholic church is not, and has never been (perhaps to the chagrin of some pontiffs), a monolithic or unanimous entity; it is composed of individuals and groups who often hold widely divergent viewpoints. This diversity of opinion was in full evidence even in the celebrated case of Galileo, where clerics and laymen are to be found distributed across the whole spectrum of responses from support to condemnation. The question, then, is what the preponderant attitude was, and in this case it is clear from the historical record that the Catholic church has been probably the largest single and longest-term patron of science in history, that many contributors to the Scientific Revolution were themselves Catholic, and that several Catholic institutions and perspectives were key influences upon the rise of modern science.[4]

In contrast to our starting myth, it is an easy matter to point to important figures of the Scientific Revolution who were themselves Catholics. The man often credited with the first major step of the Scientific Revolution, Nicolaus Copernicus (1473–1543), was not only Catholic but in Holy Orders as a cathedral canon (a cleric charged with administrative duties). And lest it be said that he was simultaneously persecuted for his astronomical work, it must be pointed out that much of his audience and support came from within the Catholic hierarchy, and especially the Papal Court (see Myth 6). His book begins with a dedication to Pope Paul III that contains an account of the various church officials who supported his work and urged its completion and publication. Galileo, too, despite his celebrated and much mythologized face-off with church officials, was and remained Catholic, and there is no reason to question the sincerity of his faith.

A catalog of Catholic contributors to the Scientific Revolution would run to many pages and exhaust the reader's patience. Thus it will suffice to mention just a very few other representatives from various scientific disciplines. In the medical sciences, there is Andreas Vesalius (1514–1564), the famous anatomist of Brussels (see Myth 5); while another Fleming, Joan Baptista Van Helmont (1579–1644), one of the most innovative and influential voices in seventeenth-century medicine and chemistry, was a devout Catholic with strong mystical leanings.[5] In Italy, the microscopist Marcello Malpighi (1628–1694) first observed capillaries, thus proving the circulation of the blood. Niels Stensen (or Nicolaus Steno, 1638–1686), who remains known today for his foundational work on fossils and the geological formation of rock strata, converted to Catholicism during his scientific work and became first a priest, then a bishop, and is currently a *beatus* (a title preliminary to official sainthood).[6] The revival and adaptation of ancient atomic ideas was due in no small part to the work of the Catholic priest Pierre Gassendi (1592–1655). The Minim friar Marin Mersenne (1588–1648), besides his own competence in mathematics, orchestrated a network of correspondence to disseminate scientific and mathematical discoveries, perhaps most notably the ideas of René Descartes (1596–1650), another Catholic.[7]

Besides individuals there are also institutions to be mentioned. The first scientific societies were organized in Italy and were financed and populated by Catholics. The earliest of these, the Accademia dei Lincei, was founded in Rome in 1603. Many other societies followed across Italy, including the Accademia del Cimento, founded in Florence in 1657, that brought together many experimentalists and former students of Galileo. Later, the Royal Academy of Sciences in Paris, founded in 1666 and probably the most stable and productive of all early scientific societies, had a majority of Catholic members, such as Gian Domenico Cassini (1625–1712), famed for his observations of Jupiter and Saturn, and Wilhelm Homberg (1653–1715), a convert to Catholicism and one of the most renowned and productive chemists of his day. Four of the early members were in orders, including the abbé Jean Picard (1620–1682), a noted astronomer, and the abbé Edme Mariotte (ca. 1620–1684), an important physicist. Even the Royal Society of London, founded in very Protestant England in 1660, had a few Catholic members, such as Sir Kenelm Digby (1603–1665), and kept up a vigorous correspondence with Catholic natural philosophers in Italy, France, and elsewhere.[8]

Catholic religious orders provided a variety of opportunities for natural-philosophical work. One of Galileo's closest early

students and supporters, and his successor to the chair of mathematics at the University of Pisa, was the Benedictine monk
Benedetto Castelli (1578–1643). But on a broader scale, during
the Scientific Revolution, Catholic monks, friars, and priests in
missions constituted a virtual worldwide web of correspondents
and data collectors. Information on local geography, flora, fauna,
mineralogy, and other subjects as well as a wealth of astronomical, meteorological, and seismological observations flooded back
into Europe from far-flung Catholic missions in the Americas,
Africa, and Asia. The data and specimens they sent back were
channeled into natural-philosophical treatises and studies by
Catholics and Protestants alike. This massive collection of new
scientific information was carried out by Franciscans, Domini-
cans, Benedictines, and, perhaps most of all, Jesuits.[9]

No account of Catholic involvement with science could be
complete without mention of the Jesuits (officially called the Society of Jesus). Formally established in 1540, the society placed such
special emphasis on education that by 1625 they had founded
nearly 450 colleges in Europe and elsewhere. Many Jesuit priests
were deeply involved in scientific issues, and many made important contributions. The reformed calendar, enacted under Pope
Gregory XIII in 1582 and still in use today, was worked out
by the Jesuit mathematician and astronomer Christoph Clavius
(1538–1612). Optics and astronomy were topics of special interest for Jesuits. Christoph Scheiner (1573–1650) studied sunspots,
Orazio Grassi (1583–1654) comets, and Giambattista Riccioli
(1598–1671) provided a star catalog, a detailed lunar map that
provided the names still used today for many of its features, and
experimentally confirmed Galileo's laws of falling bodies by measuring their exact rates of acceleration during descent. Jesuit in-
vestigators of optics and light include Francesco Maria Grimaldi
(1618–1663), who, among other things (such as collaborating
with Riccioli on the lunar map), discovered the phenomenon of
the diffraction of light and named it. Magnetism as well was studied by several Jesuits, and it was Niccolo Cabeo (1586–1650) who

devised the technique of visualizing the magnetic field lines by sprinkling iron filings on a sheet of paper laid on top of a magnet. By 1700, Jesuits held a majority of the chairs of mathematics in European universities.[10]

Undergirding such scientific activities in the early-modern period was the firm conviction that the study of nature is itself an inherently religious activity. The secrets of nature are the secrets of God. By coming to know the natural world we should, if we observe and understand rightly, come to a better understanding of their Creator. This attitude was by no means unique to Catholics, but many of the priests and other religious involved in teaching and studying natural philosophy underscored this connection. For example, the Jesuit polymath Athanasius Kircher (1602–1680) envisioned the study of magnetism not only as teaching about an invisible physical force of nature but also as providing a powerful emblem of the divine love of God that holds all creation together and draws the faithful inexorably to Him. Indeed, if Jesuit work remains today inadequately represented in accounts of scientific discovery, it is in part because science proceeded down a path of literalism and dissection rather than following the Jesuits' path of comprehensive and emblematic holism.[11]

Finally, historians of science now recognize that the impressive developments of the period called the Scientific Revolution depended in large part on positive contributions and foundations dating from the High Middle Ages, that is to say, before the origins of Protestantism.[12] This fact too must be brought to bear on the role of Catholics and their church in the Scientific Revolution. Medieval observations and theories of optics, kinematics, astronomy, matter, and other fields provided essential information and starting points for developments of the sixteenth and seventeenth centuries. The medieval establishment of universities, the development of a culture of disputation, and the logical rigor of Scholastic theology all helped to provide a climate and culture necessary for the Scientific Revolution.

Neither interest and activity in science nor criticism and suppression of its tenets align with the confessional boundary between Catholics and Protestants. Modern science is not a product of Protestantism and certainly not of atheism or agnosticism. Catholics and Protestants alike made essential and fundamental contributions to the developments of the period we have called the Scientific Revolution.

THAT RENÉ DESCARTES ORIGINATED

THE MIND-BODY DISTINCTION

Peter Harrison

This is Descartes' error: the abyssal separation between body and mind, between the sizable, dimensioned, mechanically operated, infinitely divisible body stuff, on the one hand, and the unsizable, undimensioned, unpushable, non-divisible mind stuff.

—Antonio Damasio, *Descartes' Error* (1994)

Rumor mills exist in every age. During the seventeenth century a story circulated among gentlemen of letters to the effect that René Descartes (1596–1650), now commonly identified as the father of modern philosophy, was accompanied in his travels by a life-size mechanical doll. Whether the anecdote had a factual basis or was solely the product of the malign imaginations of Descartes' enemies is not entirely clear, but certainly the story was well known.[1] On the most sympathetic interpretation, the doll was a simulacrum of Descartes' illegitimate daughter, Francine, who had tragically died when she was five. There were, needless to say, less charitable interpretations that insinuated more than sentimental attachments between Descartes and his mechanical companion.

This story is but one of a number of myths that over the years have become part and parcel of the reputation of the French philosopher. Indeed, there seems to be something about Descartes' person and his philosophy that invites slander and simplistic mischaracterization. He is, perhaps, the most maligned and misunderstood philosopher ever to have lived. Among the more common misconceptions about Descartes are these:

- Descartes was primarily a metaphysician with little interest in scientific matters.
- Insofar as he was interested in science, Descartes was an "armchair scientist" who ignored experimentation and empirical evidence.
- Descartes was a covert atheist or, at best, a deist, and his religious statements were intended to cloak his impiety.
- Descartes was a rationalist who dismissed the role of the emotions.
- Descartes was the first to posit the radical separation of mind and body, and his erroneous and unscientific dualism has been a blight on Western thought ever since.

The last of these misconceptions is perhaps the most pervasive, but it is worth briefly dispensing with the others. The current consensus among the relevant specialists is that Descartes' philosophical pursuits were secondary to his scientific interests. Moreover, while Descartes' science was perhaps less experimentally based than that of his English counterparts, he was certainly no stranger to the laboratory, and his scientific achievements were important and influential. There is no evidence that Descartes' religious convictions (Catholic) were anything but conventional and sincere. As for his supposed neglect of the emotions, as we shall see, Descartes was deeply interested in the "passions," to use the contemporary terminology, and devoted his last major work to their study.

The chief misconception about Descartes that needs to be addressed—or perhaps we should speak here of a cluster of misconceptions—is that Descartes was a thoroughgoing dualist, that his dualism involved an unbridgeable gulf between mind or soul and body, and that this erroneous and incoherent view was a disaster for Western philosophy and for attempts to understand mental functioning scientifically. These misconceptions flourish in a variety of philosophical and more popular writings. One of the more influential expressions of these views of Descartes may be found in Gilbert Ryle's classic, *The Concept of Mind* (1949), in which the Oxford philosopher derisively describes the Cartesian doctrine of mind and body as "the myth of the ghost in the machine."[2] To regard mental events as something distinct from physical events, Ryle believed, was to commit a "category mistake." Ryle's student, the philosopher Daniel Dennett, has subsequently taken it upon himself to exorcise those ghosts that survived his mentor's initial onslaught. One of Dennett's chief targets is the idea of "the Cartesian theatre"—the assumption that there is a place where thoughts and sensations come together in the brain to be observed by a single unitary consciousness. Because of Descartes' influence, he suggests elsewhere, we still tend to treat the mind as "the body's boss, the pilot of the ship."[3] Cartesian dualism, he concludes, is "fundamentally antiscientific."[4] The most recent in a long line of Descartes' detractors is the neurologist Antonio Damasio, whose popular book *Descartes' Error* (1994) leaves the reader in little doubt as to his attitude toward the seventeenth-century philosopher. The "error" of the title is identified as "the abyssal separation between body and mind." This was apparently compounded by Descartes' further separation of reason from emotion and his denial of the integration and interdependence of mind and body.[5]

From assessments such as these we can distill three widespread and closely related misreadings of Descartes. First is the assumption that Descartes was a dualist whose stance necessarily committed him to ignoring the embodied nature of human persons. Second is the view that Descartes failed to provide any account of how the distinct substances of mind and body could interact. Third is the general conclusion that the Cartesian view of mind is quasi-religious, profoundly unscientific, and philosophically unhelpful. Let us consider these in turn.

It must first be acknowledged that there are writings by Descartes that lend support to the myth, or at least part of it. It cannot be denied that in the *Meditations* and elsewhere Descartes did indeed assert that body and mind are composed of distinct substances, and it is not clear that he ever disavowed this position.[6] What is most contentious about many modern readings of Descartes on mind and body is not the claim that Descartes argued for dual substances, but rather the assumption that for Descartes this necessarily entails an "abyssal separation" (Damasio's phrase) between mind and body. In fact, Descartes took pains to deny such a separation, asserting that mind and body are "intermingled" so as to form a "unitary whole." Mind and body, he insists, form a "substantial union."[7] He also unambiguously states (*pace* Dennett) that the mind is *not* in the body "as a pilot in his ship." The pilot-ship metaphor for the relation of mind and body, incidentally, has a previous history of misattribution. The trope first appears not in Plato, with whom it is most commonly associated, but in Aristotle and subsequently in the writings of the Neoplatonist philosopher Plotinus. It is Thomas Aquinas (1255–1274) who first erroneously attributes the metaphor to Plato.[8] In fact, the doctrine of a radical separation of mind and body is one that should more properly be laid at the feet of Aristotle or Plotinus, rather than Descartes.[9]

For those who take the time to read Descartes carefully— particularly his correspondence and the later works on the passions—it is evident that the integration of mind and body, rather than their separation, increasingly preoccupied this philosopher. Descartes clearly came to the view that the best way to study this unity of mind and body that is the human being was to focus attention on the emotions or, to use the classical and early-modern category which meant something similar, the *passions*. The passions, as it turns out, play an important role in our knowledge. They are, says Descartes, kinds of perceptions or modes of knowing.[10] Descartes actually integrates the passions into the processes of knowledge in a quite new way. Far from positing a vast gulf between mind and body and writing the passions out of the equation, then, Descartes in fact asserts the opposite. So much for

Damasio's contentions about Descartes' putative "error." Indeed, in important respects, Damasio's thesis is more a realization than a refutation of the Cartesian program.[11]

Descartes' dualism, then, should not be understood as an attempt to ignore the unity of the mind and body, nor does it entail a neglect of the emotions. Indeed, the interaction of mind and body is so central a concern for Descartes that some commentators have gone so far as to suggest that it is misleading to refer to Descartes as a dualist.[12] Certainly it seems that Descartes was committed to understanding the world in terms of not two but three basic kinds of entity—extended material things (matter), thinking things (minds), and mind-body composites (persons).[13] Leading Descartes scholar John Cottingham has thus suggested an alternative way of characterizing the Cartesian position—not "dualism" but "trialism."[14]

This brings us to the second element of the myth—that the Cartesian position cannot explain how mind and body interact. The problem is straightforward: how does the mind cause the body to move, and how do the bodily senses cause conscious states, given that mind and body are distinct substances? Undergraduates are often taught either that Descartes failed to provide an adequate account of these interactions or that he and subsequent Cartesians sought refuge in an ad hoc and deeply implausible thesis called "occasionalism." This is the idea that there is in fact no real causal interaction between mind and body—that when I form a conscious intent to move my arm, for example, God routinely provides the necessary connection by directly causing my arm to move. While occasionalism does seem to provide a solution to the problem of the correlation of our mental intentions with bodily movements, it was not developed for that purpose. Rather, it arose out of concerns to address a more general difficulty with causation: how, for example, do the inert particles of matter posited by the mechanical philosophy exert any causal influence on each other? On the occasionalist account, matter-matter causation is as problematic as mind-matter causation. This skepticism about natural causation arose as part of the rejection of the Aristotelian understanding of causation, and it finds its logical terminus in David Hume's (1711–1776) contention that what we speak of as causes and effects are really just events that we observe to be constantly conjoined.[15]

Again, however, there is an element of truth in the occasionalism myth. Although occasionalism was not developed with a view to solving the problem of mind-body interactions—this was an added bonus—it nonetheless highlights an alternative way of approaching this apparently intractable problem: to see it as arising out of a deficiency in our conception of causal explanation. On one plausible reading, Descartes asserts that the correlations between mental events and bodily movements are simply natural properties of the body-mind amalgam. In much the same way that God established the physical laws that govern the interactions of material things—the laws of nature—God also decreed what correlations would obtain between mental events and bodily events. The relations of mind and body, on this account, are explained in terms of psychophysical laws that constitute our very nature as embodied beings. To seek an explanation of the operations of the mind-body composite along the lines of the kinds of relations that obtain among material bodies or ideas is, ironically, to commit a kind of category mistake by searching for the wrong kind of explanation for a primitive state of affairs.[16]

As for the supposedly "unscientific" character of Descartes' view of the relation of mind and body, this claim involves a profoundly ahistorical assessment of Descartes' achievements. A key implication of Descartes' theory of mind was that the physical world was to be understood as passive matter. This view played a crucial role in banishing from nature the quasi-spiritual "forms" of Aristotelianism. For this reason the chief complaint about Cartesian philosophy during the seventeenth and eighteenth centuries was not its dualism but its incipient materialism. The point, then, is not so much that Descartes installed a ghost in the human machine, to hark back to Ryle's epigram, but that he successfully banished "ghosts" from the rest of the material world. In so doing he made space for a whole new range of what we would call "scientific" explanations. These explanations, as we have already noted, were couched in terms of "laws" rather than "causes" and thus played an important role in establishing the principles of modern science.

In sum, Descartes' views about mind, body, and their relation are subtle, sophisticated, and complex. They bear little resemblance to the simplistic caricatures that often pose as authoritative accounts of his work. Descartes gave a central place to the emotions in his psychology, and he took very seriously the embodied nature of human beings. Because of Descartes' insistence that the mind-body amalgam was a real entity, some commentators have gone so far as to suggest that he no longer be numbered in the ranks of the dualists.

It is worth asking, in conclusion, why this myth has proven so persistent. Certainly, like all myths, it contains a kernel of truth. More importantly, though, narratives about "Descartes' myth" (Ryle), the "Cartesian Theatre" (Dennett), or "Descartes' Error" (Damasio) provide an impressive historical backdrop against which contemporary thinkers can outline their own theories of the human mind. The compelling plot line is that at a certain point in its history, Western thinking about the mind or soul took a wrong turn and that we have labored under multiple confusions ever since. Those who recount this now familiar story about Descartes' role in our present misfortunes, by implication, present themselves as our philosophical saviors, offering solutions that will set us once again on the right path. The magnitude of their achievement is gauged not only by the logic of their arguments but by the stature of the philosophical giant they have slain. Small wonder, then, that a celebrated figure such as Descartes should so frequently be enlisted by those seeking to establish themselves as iconoclastic visionaries in the field of the philosophy of mind.

Finally, we should relate this myth to more general discussions within the field of science and religion. The position mistakenly attributed to Descartes is also typically assumed to be more or less the Christian understanding of persons. (Ryle contended, for example, that Descartes was simply reformulating a prevalent theological doctrine of the soul.)[17] Accordingly, scientifically motivated critics of the so-called Cartesian view of the mind often regard Descartes as having underwritten an essentially religious position; they imagine that in attacking him they are inflicting collateral damage on a basic tenet of religious belief. Again, however, these criticisms miss their mark, for they underestimate the value placed on the physical world and on the embodiment of human persons in Western religious traditions. Although this will come as a surprise to some, orthodox Christianity (in contrast to Platonism and gnosticism) assumes a holistic view of the person and a positive view of embodiment—so much so that even in the next life souls will be reunited with a resurrected body. The doctrine of an abyssal separation of body and soul was not propounded by Descartes, and neither is it a tenet of Christian belief.

THAT ISAAC NEWTON'S MECHANISTIC COSMOLOGY ELIMINATED THE NEED FOR GOD

Edward B. Davis

The Clockwork Universe Theory is a theory, established by Isaac Newton, as to the origins of the universe.

A "clockwork universe" can be thought of as being a clock wound up by God and ticking along, as a perfect machine, with its gears governed by the laws of physics.

What sets this theory apart from others is the idea that God's only contribution to the universe was to set everything in motion, and from there the laws of science took hold and have governed every sequence of events since that time.

—Wikipedia, the free encyclopedia

The metaphor of the mechanical clock in Newton's construction of the heavens and its legacy illustrate the power of metaphors in the development of scientific thought.

—Sylvan S. Schweber, "John Herschel and Charles Darwin" (1989)

With Aristotle's laws of motion overthrown, no role remained for a Prime Mover, or for Moving Spirits. The hand of God, which once kept the heavenly bodies in their orbits, had been replaced by universal gravitation. Miracles had no place in a system whose workings were automatic and unvarying. Governed by precise mathematical and mechanical laws, Newton's universe seemed capable of running itself.

—Thomas H. Greer, *A Brief History of the Western World* (1982)

The typical picture of Isaac Newton (1642–1727) as the paragon of Enlightenment deism—responsible for recasting God as a divine clockmaker with nothing more to do once he had completed his creation—is more than just badly mistaken: it is precisely the opposite of the truth. It cannot simply be corrected; it must be utterly repudiated. In fact, Newton rejected both the clockwork metaphor itself and the cold mechanical universe upon which it is based. His conception of the world entailed a deep commitment to the constant activity of the divine will, unencumbered by the deists' "rational" restrictions that later transformed the sovereign ruler of the universe into a mere constitutional monarch who cannot violate his own "laws."[1]

It has long been known that Newton had serious interests in theology, biblical prophecy, church history, and alchemy, to which he committed many years of his life and thousands of sheets of paper. Until very recently, however, scholars have generally denied that the enormous energy he devoted to those pursuits had any discernible effect on his scientific work—except unfortunately to take significant amounts of time away from it. Newton, so the standard story goes, was a great physicist and mathematician who rather embarrassingly "dabbled" in alchemy and theology; if he looked over his shoulder at the "dark ages" a little too often and too intently, it was mainly a result of the nervous breakdown he suffered in the fifty-first year of his life, after which he didn't do very much science anyway.

Two things have led most contemporary scholars to draw very different conclusions. First, Newton's voluminous alchemical and theological manuscripts, now scattered over three continents, have been scrutinized by a diverse community of specialists who have studied both the papers and their historical contexts much more fully than previously. Added to this new knowledge is something even more important: a new historical attitude that takes at face value what Newton actually said and did, without imposing our modern cultural norms and beliefs on him and his milieu. History fails to be reliable whenever it neglects to show us the world as it looked to the historical actors themselves—and that is exactly where the traditional view of Newton went so badly wrong.

When viewed through his own writings, many of them neglected or never really understood until recently, Newton is seen as a deeply pious and serious student of theology whose ideas about God and the Bible helped to shape his whole view of the world, including his conception of nature and how it works. Starting when he was about thirty years old, Newton began wide-ranging and painstakingly detailed investigations of theology and church history, focusing especially on the doctrine of the Trinity. Having read all of the important patristic writers himself and scrutinized biblical passages, he emphasized the authenticity of those texts that speak of the Son's subordination to the Father while dismissing other texts typically used to support the Trinity (including 1 John 5:7 and 1 Timothy 3:16) as later "corruptions" of Scripture. Soon he concluded that Christ was the Son of God and preexistent before creation, but not co-eternal with and equal to God the Father. The created Word (λόγος) made flesh, Christ should be worshiped for his obedience unto death—for what he had done, not for who he is. Though a divine mediator, Christ was subordinate to the Father, whose will he carried out.

Newton's precise beliefs on certain points and how he arrived at them are still debated by scholars. It is not entirely clear whether he is best understood as an Arian (as most scholars think) or as a Socinian or some other type of anti-Trinitarian, but it is transparently clear that he saw the Trinity as a false and idolatrous doctrine, an abomination that had been criminally foisted on the church in the fourth century by a deceitful Athanasius (ca. 293–373), bishop of Alexandria. In seventeenth-century England such heterodox convictions were not tolerated, so Newton shared them with only a few carefully chosen men of like mind, among them Samuel Clarke (1675–1729) and William Whiston (1667–1752). Nevertheless, veiled hints of his heretical Christology appear in a few prominent publications, such as the "General Scholium" to the second (1713) edition of his magnum opus, the *Principia*. Overall, he regarded himself as one of just a tiny remnant of true believers in a pristine, properly monotheistic Christianity that had to be preserved and would eventually be restored.

It is equally clear that Newton was no deist, despite the fact that his greatest biographer, the late Richard S. Westfall (1924–1996), repeatedly stated it as an obvious point—a most ironic fact, given that no one had ever done more to publicize Newton's devotion to theological study. Like many of the best scholars of his generation, Westfall saw Newton through strongly tinted modernist lenses, and therefore he ultimately misunderstood a central feature of Newton's religion. According to Westfall, Newton elevated reason over Scripture and denied the Trinity because he found it too mysterious and irrational; rationalism in religion thus made Newton a protodeist, in Westfall's judgment. In reality, unlike the deists, Newton trusted the Bible (except for its Trinitarian "corruptions") and often took it literally, especially prophetic texts from Daniel and Revelation. He believed in predestination, the bodily resurrection of Jesus, the future resurrection of the faithful, and the millennial kingdom ruled by Christ. The biblical text itself held priority over "deductions" that reason might draw out of the text—a primary example being the erroneous doctrine of the Trinity. In short, Newton denied the Trinity because in his view the uncorrupted Bible does not teach it, not because reason led him to deny a genuine biblical doctrine.[2]

Roughly four years before he plunged deeply into theology, Newton had dived into even murkier waters—in which he remained submerged for at least three decades. The very word *alchemy* suggests to modern ears the same things it suggested to many ears in the mid-eighteenth century, shortly after Newton's death: secret formulas, magic, and fruitless efforts to turn lead into gold. However, for Newton and some of his most illustrious contemporaries, including Robert Boyle (1627–1691) and John Locke (1632–1704), alchemy was a very serious part of chemistry that held great promise for understanding the nature of matter. Very possibly, Newton even considered alchemy a way in which to probe the profound mystery of how God worked through intermediate agents to carry out his will in the natural world.[3] We should not wonder, then, why he penned more than one million words on alchemical subjects.

Newton's conviction that God governed the world actively and constantly, but usually indirectly, probably derived from the Arian emphasis on Christ as God's agent in creating the world. In one of his unpublished manuscripts, Newton wrote that Christ

> was in the beginning with God: All things were made by him & without him was nothing made that was made . . . As Christ is now gone to prepare a better place for the blessed so in the beginning he prepared this place for mortals being in glory with the father before 1 John. For the supreme God doth nothing by himself which he can do by others.[4]

As God's viceroy, Christ controlled the "active principles" that brought together particles of inert matter to form the various bodies, living and nonliving, that compose the universe. Newton was well read in René Descartes (1596–1650) and other mechanical philosophers, who sought to explain phenomena in terms of matter and motion. He was a mechanical philosopher himself, but he found Descartes' specific picture of the world as nothing but matter and motion theologically dangerous—where was there any room for free divine activity? Alchemy offered the answer: matter in its essence was incapable of holding itself together or influencing other matter except by direct contact, so the forces and powers manifested in chemical phenomena bore witness to the mediated activity of the creator, who had made matter from nothing in the beginning and who could move it around now at his volition.

Newton's understanding of the cosmic force of gravitation
was similar to his understanding of alchemy. Mathematical
analyses of motion in the heavens convinced Newton of the reality of attractive forces operating between pieces of matter, but
when he put forth his full physical theory in the first edition of
the *Principia* (1687), he said nothing about gravity's cause. A
few years earlier, he had abandoned the idea that a mechanical
ether, filling all of space, could explain gravitation; there was no
way the ether could do that without also hindering the motions
of the planets in ways that contradicted observations. Shortly after the *Principia* was published, however, Newton came to believe that God the Father might be the direct, immediate cause
of gravitation: the omnipresent God, in whom we live and move
and have our being, moves matter through space—what Newton
called God's "sensorium"—as readily as we move the parts of
our own bodies. Queries 28 and 31 in the *Opticks* (written for
the Latin translation of 1706) and parts of the General Scholium
reflect this view, although the theology embedded within those
texts is missed by most readers.[5]

Newton's thinking in the years surrounding the publication of
the *Principia* was driven by an overriding belief in the importance
of God's dominion, the reality of which he believed natural philosophy could demonstrate from studies of phenomena here on
earth and in the heavens. Dominion was indeed the defining characteristic of his God: "a being, however perfect, without dominion is not the Lord God," he wrote in the General Scholium.[6]
Consequently, Newton flatly refused to speak of the world as a
clockwork that runs on its own, without any need for ongoing divine governance. When the German philosopher Gottfried Leibniz (1646–1716) questioned Newton's belief that God may need
periodically to adjust the motions of the planets, in order to prevent the universe from running down, it was he—not Newton—
who brought clocks explicitly into the conversation. If God had
"to wind up his watch from time to time," Leibniz claimed, then
he lacked "sufficient foresight to make it a perpetual motion,"

obliging him "to clean it now and then by an extraordinary con-
course, and even to mend it, as a clockmaker mends his work,"
casting doubt on the skill of the divine craftsman, who "is oftener
obliged to mend his work and set it right." In the reply delivered
by his protégé Samuel Clarke, Newton explicitly rejected the
clockwork universe that is so often wrongly associated with his
name: "The notion of the world's being a great machine, going
on without the interposition of God, as a clock continues to go
without the assistance of a clockmaker; is the notion of materialism and fate, and tends, (under pretence of making God a *supra-
mundane intelligence,*) to exclude providence and God's
government in reality out of the world."[7]

Manifestly, Newton's God was no Enlightenment absentee
clockmaker. Rather he was free to make a world of any sort he
pleased, and if he chose to alter it later, that was the prerogative
of an omnipotent, providential governor who exercises his dominion over all that comes to pass—who are we mere mortals
to question his foresight? Although the clockmaker God is frequently associated with "Newtonianism," Newton was not a
Newtonian in this sense. How did he come to be seen as the Enlightenment man that he never was? Partly because his theological and alchemical papers were closely guarded secrets during
his own life, and before the latter part of the twentieth century
scholars did not take them as seriously as they should have. Furthermore, the French *philosophes* of the eighteenth century created their own view of Newton as the apotheosis of the kind of
secular reason that they advanced to replace Christianity, and an
active God did not fit into their picture. At the same time, the
practical application of Newton's physics using force, inertia,
and fixed "laws" to account for motion in the heavens and on
the earth did not require the invocation of divine agency as part
of the explanation; the laws alone seemed sufficient for the job,
provided that ultimate questions were left to one side.

Newton's physics is rarely taught today alongside the metaphysical and theological conceptions with which it was inextricably

linked in his own mind. This is probably less true of Albert Einstein's theory of relativity or Charles Darwin's theory of evolution—a modernist bias once again rears its head. If we are going to separate a scientific theory from the larger intellectual context in which it arose, we must be especially careful not to associate its founder with views directly opposed to those he or she actually held. It is high time that Newton's clock stopped ticking.

THAT THE CHURCH DENOUNCED ANESTHESIA
IN CHILDBIRTH ON BIBLICAL GROUNDS

Rennie B. Schoepflin

From pulpit after pulpit Simpson's use of chloroform was
denounced as impious and contrary to Holy Writ; texts were
cited abundantly, the ordinary declaration being that to use
chloroform was "to avoid one part of the primeval curse on
woman."

—Andrew Dickson White, *A History of the Warfare
of Science with Theology in Christendom* (1896)

When 19th-century doctors began using chloroform to alleviate
the pain of childbirth, the Scottish Calvinist church declared it a
"Satanic invention" intended to frustrate the Lord's design.

—Deborah Blum, *New York Times* (2006)

Among the evidence cited to illustrate the ways in which a back-
ward church has delayed the advances of science is the persistent
myth that organized Christianity opposed the use of anesthesia
in childbirth. The book of Genesis relates the initial facts of
childbearing: As punishment for their sin, God cursed Adam,
Eve, and their descendants; thereafter, men would plant and har-
vest by the sweat of their brows and women bear children in
pain and suffering. Supposedly read by church authorities and
pious believers as a divine command for all eternity, this passage
proscribed the administration during labor of anesthesia, whose
use revealed a defiant effort by rebellious humans to subvert

God's purpose. For the church and her believers, such scientific impiety, "contrary to Holy Writ," could only come from the Prince of Demons as a "Satanic invention" to overthrow the will of God. For defenders of the advances of science, such religious obscurantism only confirmed their suspicions about the church's threat to human progress. But is this the whole story, and if not, why has it persisted, repeated by caviler, believer, and expert alike for over one hundred years?

Soon after the discovery of ether-based anesthesia in 1846 the influential Edinburgh obstetrician James Young Simpson (1811–1870) quickly adopted it for relieving pain during delivery.[1] After Simpson discovered chloroform's anesthetic qualities in late 1847, he established himself as a champion for the control of pain in childbirth by tirelessly arguing for its safety and efficacy against scientific opposition and defending its use against religious and moral objections. Already by December 1847 he had published a pamphlet entitled "Answer to the Religious Objections Advanced against the Employment of Anaesthetic Agents in Midwifery and Surgery." Directed primarily to medical professionals, this pamphlet argued on exegetical, logical, historical, and moral grounds against those who "believe that the practice in question ought in any degree to be opposed and rejected on religious grounds."[2] Typical of many of his medical and scientific contemporaries, Simpson denied any inherent conflict between religion and science, and in an 1848 letter to London's Dr. Protheroe Smith he affirmed "that the language of the Bible is, on this as on other points, strictly and scientifically correct."[3]

Theologians and clergy from Presbyterian, Anglican, and various dissenting churches sent overwhelmingly positive responses to Simpson for his 1847 pamphlet. Nonetheless, religious objections surfaced among some of their colleagues and among laypersons.[4] Simpson noted that in Edinburgh a "few among the clergy themselves, for a time, joined in the cry against the new practice" but that by mid-1848 he no longer met "with any objections on this point, for the religious, like the other forms of opposition to chloroform, have ceased among us."[5] The Reverend Thomas Boodle of Surrey read Simpson's pamphlet and reported that it had "relieved my mind from the serious [religious?] objections I had entertained" but that he desired further information regarding its "safety and expediency" for midwifery.[6] And Robert Gaye, clergyman in the Established Church of Northern Ireland, corresponded similarly that "it may be gratifying to you to know that my poor opinion and the general one here among all persons who have perused your work clerical and medical—is, that you have written a deeply interesting pamphlet on a subject about which we doubted almost if any sane person could have a second opinion."[7] Simpson may have been guilty of hyperbole in declaring that all opposition to chloroform had ceased by mid-1848; some evidence suggests that it lingered until Queen Victoria delivered Prince Leopold under anesthesia in 1853.[8] Nonetheless, no evidence supports the notion that the opposition was widespread or orchestrated by organized Christianity.

A. D. Farr, a historian and physician who has conducted an exhaustive study of the matter, found only fleeting published evidence "either for theological opposition to anaesthesia from the institutional churches or of any widely held (or express) opposition on the part of individuals." He concluded that "it is almost certain that Simpson's pamphlet . . . was written to forestall objections which, in the event, did not arise, and that its publication has subsequently been mis-interpreted by other commentators as evidence for a non-existent opposition."[9] Whether one agrees with Farr's conclusion regarding Simpson's forethought or not, he was doubtless correct that organized religion in the United Kingdom mounted no formal attack on the use of anesthesia in childbirth.

In the spring of 1847 Frances "Fanny" Appleton Longfellow (1817–1861), the second wife of Henry Wadsworth Longfellow (1807–1882), became the first American woman to receive ether during childbirth, and by January 1848 Simpson's "Answer to the Religious Objections" had arrived in Boston.[10] Harvard physician and obstetrics professor Walter Channing (1786–1876) quickly embraced the use of ether and chloroform to control the pain and discomforts of childbirth in 1848, arguing forcefully for their safety and efficacy. In his widely read *Treatise on Etherization in Childbirth* (1848), Channing presented the medical and scientific case for the administration of anesthesia in labor and refuted current religious and moral objections.[11]

Channing began his defense by enumerating examples of the way in which the supposed teachings of the Bible had been misused in the past to justify all sorts of misguided and contradictory beliefs and behaviors: a stationary earth, warfare, both the total abstinence from and moderate use of alcohol, capital punishment, and blind submission to government. "And, finally, etherization has, with some," he continued, "its insuperable obstacle in the third chapter of Genesis." He had heard that some Christians objected to the use of anesthesia in labor on the basis of Genesis 3:16, and he knew of Simpson's pamphlet, but he had assumed that such views were "confined very much to the clergy." But on closer examination he discovered "that the people were receiving the doctrine, and that medical men were among its advocates."[12]

Channing wrote to George Rapall Noyes (1798–1868), Harvard professor of Hebrew and Oriental Languages, to solicit his interpretation of the Genesis passage on pain in childbirth. Noyes replied that "I should as soon believe, that labor-saving machines were in opposition to the declaration, 'In the sweat of thy face shalt thou eat bread;' or that the cultivation and clearing of land was opposed to the declaration, 'Thorns also and thistles shall it bring forth to thee' " as believe that Genesis forbade the use of anesthesia to relieve the pain of childbirth. For Channing this put to sleep any justifiable biblical reason to avoid anesthesia in childbirth, but as in the case of Edinburgh, "The interest of our subject has extended beyond the medical profession, and has even reached the pulpit." Channing related an anecdote about a minister who had recently preached a sermon titled "Deliver Us from Evil" on the dangers of anesthesia in childbirth. "Said one to a friend, as he left the church, 'How did you like our sermon?'— 'Very well' was the reply. 'It is not wholly wrong to lessen or destroy pain. *We may eat peppermints!*' "[13]

After the Civil War some organized opposition to the use of anesthesia in childbirth may have surfaced in the United States, as revealed by the fact that the American Medical Association in 1888 felt it necessary to dismiss "religious objections to obstetric anesthesia as 'absurd and futile.'"[14] But as in the United Kingdom, little or no evidence supports the claim that the church mounted a systematic or sustained attack; to the contrary, the record reveals that much of the religious and moral opposition arose among medical professionals themselves. As historian Sylvia D. Hoffert concluded,

> There is no evidence to indicate that parturient women or members of their families were in the least concerned with any of the religious, philosophical, or social issues being debated by members of the medical community. Their own experience or what they knew of the experience of others told them that they could expect to suffer during labor. They were quite willing to try something that might relieve them in their anticipated agony.[15]

Not every physician was as confident of the safety and efficacy of anesthesia as Simpson and Channing or, as we have seen, as worry-free about its religious implications. But contrary to the oft-repeated view that church authorities spearheaded opposition to anesthesia in childbirth, obstetrical experts—from Ireland and England to France, Germany, and the United States—proved to be its strongest opponents, but on medical, not religious, grounds.[16] Charles D. Meigs (1792–1869), professor of obstetrics at Jefferson Medical College in Philadelphia for over forty years, proved to be one of its most vocal and influential opponents. But even Meigs's objections have been misinterpreted by later commentators to have been founded primarily on religion and morality, not science.[17]

Meigs believed "etherization in Midwifery" to be both "unnecessary" and "improper" and a contravention of "the operation of those natural and physiological forces that the Divinity
has ordained us to enjoy or to suffer." Natural law and not biblical injunction or church authority, however, informed Meigs's
objections. Rather than question the beneficence of God in or-
daining "woman to the sorrow and pain of them that travail in
childbirth," we ought, according to Meigs, acknowledge that an
"economical connection exists betwixt the power and the pain
of labors. While, therefore, we may assume the privilege to control, check and diminish the pains of labor whenever they become so great as to be properly deemed pathological," he wrote,
"I deny that we have the professional right, in order to prevent
or obviate them, to place the lives of women on the hazard of
that *progress* of anaesthesia, whose laws are not, and probably
can never be ascertained so as to be foreknown."[18] For Meigs,
the force necessary to free the infant from the mother was inseparable from the resultant pain; the pain was due not to a biblical
"curse" but rather to the natural, physical results of mechanical
force on living tissue. And when they did not interfere with the
necessary pulses of force necessary for a natural childbirth, pain-
control measures should be applied.

When Meigs referred to the "profound *Drunkenness* of etherization" and asserted that "to be insensible from whisky, and gin,
and brandy, and wine, and beer, and ether, and chloroform, is to
be what in the world is called Dead-drunk," his concerns were
not moral but medical. Meigs feared that women would not be
conscious enough under the influence of anesthesia to respond to
physicians' queries during delivery—when, for example, they had
to manipulate the infant with forceps. Echoing the ancient Hip-
pocratic mandate to "do no harm," he concluded that natural labor "is the culminating point of the female somatic forces. There
is, in natural labor, no element of disease—and, therefore, the
good old writers have said nothing truer nor wiser than their old
saying, that '*a meddlesome midwifery is bad.*'"[19]

Medical historian Martin S. Pernick's study of shifting American attitudes toward pain and its alleviation provides essential context for understanding these reactions to anesthesia and childbirth. He found that "a wide variety of nineteenth-century healers concluded that the pains of childbirth and disease were deserved punishments, chastisements that might be immoral and unhealthy to anesthetize away." Although he concluded that "a literal reading of the Bible did play a role in such opinion," he also found that "such arguments were derived more from the doctrines of natural healing than from the book of Genesis. And the most extreme exponents of such views were not strict predestinarians but such radical perfectionists as the hydropaths and Grahamites."[20]

Despite the overwhelming evidence that nineteenth-century physicians—not clergy—drove opposition to anesthesia in childbirth and that their objections centered on medical and not religious grounds, the contrary view persists. Why have such false conclusions continued? A central reason lies in the ongoing and unwarranted use of two nineteenth-century polemics as scholarly authorities: John William Draper's *History of the Conflict between Religion and Science* (1874) and Andrew Dickson White's *A History of the Warfare of Science with Theology in Christendom* (1896). Both authors repeated the claim that the use of anesthesia in labor was discouraged not so much for physiological as for biblical reasons and the fear of impiety.[21] Bertrand Russell picked up the refrain in *Religion and Science* (1935), suggesting that "another occasion for theological intervention to prevent the mitigation of human suffering was the discovery of anaesthetics."[22] Perpetuating the myth, pathologist and medical writer Thomas Dormandy asserted in *The Worst of Evils: The Fight against Pain* (2006) that in the struggle over the use of chloroform obstetricians raised "comparatively mild objections" to its use, and they raised louder objections on "moral and religious grounds." But physician objections paled "compared to the deep religious misgivings expressed both by ministers of the Kirk and by devout layman [sic]."[23]

Another reason lies in the decentralized nature of belief and practice in the modern world. Contrary to the monolithic images of science and religion promoted by White and Draper, more often than not the effectiveness of church authority follows rather than determines constituent belief and behavior. In this case it was a scattering of laypersons, not medical or religious authorities, who raised concerns about a loss of moral control while under sedation and who encouraged a belief in the virtue of suffering through childbirth.

In the fall of 1956 Pope Pius XII (1876–1958), responding to concerns raised by the Italian Society of the Science of Anesthetics, asserted that a doctor who uses anesthesia in his practice "enters into contradiction neither with the natural moral order nor with the specifically Christian ideal"; patients "desirous of avoiding or soothing the pain can, without disquiet of conscience, make use of the means discovered by science and which, in themselves, are not immoral. Particular circumstances can impose another line of conduct, but the Christian's duty of renunciation and of interior purification is not an obstacle to the use of anesthetics."[24] This clear statement from arguably one of the most conservative Christian churches reveals the continued bankruptcy of the myth that the church denounced the use of anesthesia in childbirth on biblical grounds.

THAT THE THEORY OF ORGANIC EVOLUTION
IS BASED ON CIRCULAR REASONING

Nicolaas A. Rupke

Creationists have long insisted that the main evidence for
evolution—the fossil record—involves a serious case of circular
reasoning.

> —Henry M. Morris, "Circular Reasoning in Evolutionary
> Biology" (1977)

Modern Darwinists continue to use homology as evidence for
their theory. But . . . if homology is *defined* as similarity due
to common descent, then it is circular reasoning to use it as
evidence for common descent.

> —Jonathan Wells, *Icons of Evolution: Science
> or Myth?* (2000)

Can the Darwinian theory of evolution be dismissed as mainly a
case of circular reasoning? Does evolutionary science resemble a
silly dog chasing its own tail, excitedly circling but never getting
what it wants—namely, the factual evidence it sorely lacks?
Many creationists think so. Some of them may prefer another
metaphor to dismiss Darwinian theory, such as a serpent biting
its own tail, used by the staunch anti-Darwinian writer and
creator of the priest-detective Father Brown, G. K. Chesterton
(1874–1936). In *The Everlasting Man* (1925), Chesterton used
this image as a symbol of what he thought was the circular, self-
defeating reasoning of much of non-Christian philosophy.[1] But

whatever metaphor they choose, creationists consider the logical foundations of evolutionary theory fatally flawed.

Traditionally, the two most important scientific fields to provide evidence for organic evolution have been geology and biology. The first gave us the geologic column, also known as the stratigraphic table, which shows the global succession from bottom to top and from ancient to recent of the rock formations that over many millions of years have come to compose the crust of the earth. The geologic column additionally exhibits the fossil record, the remnants of past forms of life that for the most part have become extinct. This record reveals a progressive trend from simple to complex, "from monad to man." The first book in the English language to espouse organic evolution, *Vestiges of the Natural History of Creation* (1844)—anonymously printed but written by the Edinburgh publisher Robert Chambers (1802–1871)—almost entirely relied on the long and progressive history of the earth and life as indicated by the geologic column. Charles Darwin (1809–1882), too, although more circumspectly than Chambers, had recourse to the fossil record in his *On the Origin of Species* (1859) when trying to validate his theory of descent with modification.

Moreover, Darwin appealed to such phenomena of biology as "the unity of type," the fact that all species belonging to, for example, the vertebrates—animals with backbones and spinal cords—are built on a common architectural plan. An organ, such as a forelimb, in one vertebrate species can be traced to a similar organ in all other vertebrate species, where it occurs in exactly the same relative position, even though it may have a different shape or function. Such similarities are called *homologies*. An arm in a human, for instance, is the homologue of a fin in a fish. Homological continuities from one species to another, Darwin claimed, beautifully fitted his theory of the evolution of all species from a common ancestor.

"Not so fast," say twentieth- and twenty-first-century creationists. Strict logic shows, they insist, that the arguments for evolution from the stratigraphic column as well as from homology are invalid, because they represent instances of crude circular reasoning. Put simply, these proofs of evolution are based on the assumption of evolution. This objection was put forward most prominently by George McCready Price (1870–1963), the Seventh-day Adventist who founded "flood geology" (named for the biblical account of Noah and the flood). Price alleged in *The New Geology* (1923) that the geologic column is an artifact cobbled together on the basis of the a priori belief in an evolutionary progression of life through time.[2] Price's argument against evolution gained popularity with the creation science movement in the 1960s, which took off in the wake of what became its canonical text, *The Genesis Flood* (1961). This book, written by a conservative evangelical Old Testament scholar, John C. Whitcomb (b. 1924), and a Southern Baptist professor of hydraulic engineering, Henry M. Morris (1918–2006), significantly enlarged upon Price's young-earth creationism, which attributed the formation of nearly the entire sedimentary crust of the earth to a single catastrophic event, the biblical deluge.

Flood geology made a nonsense of the standard stratigraphic table: its proponents argued that it showed neither a long history of gradual rock accumulation nor an orderly succession of fossil progress. Morris, the founder and first president of the Institute for Creation Research, maintained—as had Price—that the apparent progression of fossils in the geologic column, adduced as proof of evolution, was an artifice produced by geologists who espoused evolution in the first place and used it to date rock formations in arranging the stratigraphic table. "This is obviously circular reasoning."[3] Given the fact that "the only genuine historical evidence for the truth of evolution is found in this fossils record," evolution theory collapses like a house of cards.[4] Morris and Whitcomb summed up the case: "The importance of the fossils in the dating of the geologic strata cannot be overemphasized. It is remarkable that the vicious circle of reasoning in this procedure cannot be appreciated by paleontologists. The *fossils alone* are used to assign a geologic time to the rock stratum, and yet this very sequence of fossils is said to constitute the greatest proof of organic evolution!"[5]

The allegations of circular reasoning have become part of the young-earth creationists' stock-in-trade against Darwin and evolution.[6] Moreover, this anti-Darwinian line of argument has been adopted by the more recent intelligent design (ID) movement, centered on the Discovery Institute. Proponents of ID habitually avoid addressing the issue of the age of the earth or the validity of the geologic column. Yet they, too, fight Darwinism and neo-Darwinist views by pointing to what they believe are evolution's fatal weaknesses of logical argumentation. Among them is Jonathan Wells (b. 1942), biologist, theologian, and minister in the Unification Church, educated at Berkeley and Yale, and a fellow of the Discovery Institute's Center for Science and Culture.[7] In *Icons of Evolution* (2000), Wells discusses "why much of what we teach about evolution is wrong." One of the "icons" he cites, conventionally used as proof of evolution, is the phenomenon of homology. "But what precisely is homology?" Wells asks. Under the heading "Homology and circular reasoning," he explains that evolutionary biologists define the term as the similarity between different species that is due to their shared ancestry. In other words, homology indicates evolution and evolution produces homology—a perfect "circular argument."

> Consider the example of bone patterns in forelimbs, which Darwin regarded as evidence for the common ancestry of the vertebrates. A neo-Darwinist who wants to determine whether vertebrate forelimbs are homologous must first determine whether they [the species being compared] are derived from a common ancestor. In other words, there must be evidence for common ancestry before limbs can be called homologous. But then to turn around and argue that homologous limbs point to common ancestry is a vicious circle: Common ancestry demonstrates homology which demonstrates common ancestry.[8]

It becomes no easier for the Darwinists—Wells maintains—when they turn to the fossil record to help determine evolutionary relationships. "Unfortunately, comparing fossils is no more straightforward than comparing live specimens . . . Any attempt to infer evolutionary relationships among fossils based on homology-as-common-ancestry 'soon leads to a tangle of circular arguments from which there is no escape.' "[9]

Some evolutionists have talked back in an effort to absolve themselves from the sins of chasing their tails.[10] But this has proved no easy matter. Creation scientists aren't fools and, strictly speaking, do have logic on their side in the major cases cited here. Moreover, they have been able to amass quotations from concerned evolutionists who apprehensively admit to the illogical practices of which they stand accused.[11] Quite likely, a variety of paleontologico-stratigraphical studies are unreliable because they are founded on a *petitio principii* (the logical fallacy of "begging the question").

All the same, the assertion that evolution is crucially based on circular arguments is a myth. There is a simple reason why the use of fossil progress and of homology in support of the theory of evolution does not constitute circular argument, and this reason is provided by history. Both the geologic column and the theory of vertebrate homology acquired their more or less definitive shape some ten to twenty years before Darwin's *Origin of Species* broke upon the world, and even a few years before the *Vestiges* created a "Victorian sensation." Neither field of evidence was, in its formative stages, driven by a belief in Chambers's species transmutation or Darwin's descent with modification. At the time of their inception, the stratigraphic table and homology had nothing whatsoever to do with evolution. Not only were they unrelated to Darwinism or any of its precursor hypotheses about the origin of species but, what's more, both reached their mature form in a context of creationist science, albeit old-earth creationism. (At least this was the case in the English-speaking world, on which I here focus attention. In Germany and France,

developments were somewhat different, although not fundamentally different with respect to the initial lack of importance
of evolution theory.)[12]

Let me provide historical specifics, first with respect to the geologic column.[13] In Britain, as early as about 1820 the outline of
a stratigraphic table existed, a skeleton table from which the
later, essentially complete version of approximately 1840 grew
by gradual additions and partial corrections. The Anglican clergyman and Oxford geologist William Buckland (1784–1856)
produced this early table of rock formations. The lineage of
Buckland's skeleton column extended back to the 1790s. During
that decade a school of mineralogical stratigraphy originated
with the work of the Lutheran mineralogist Abraham Gottlob
Werner (1749–1817), who taught at the Freiberg Mining Academy in Saxony. Interest in Werner's ideas spread across Europe,
reaching among other places Edinburgh. On the basis of the
physical superposition and mineralogical composition of the
strata, Werner recognized four major successive formations,
which he interpreted as periods of earth history.

During the first decade of the nineteenth century, the French
Lutheran paleontologist Georges Cuvier (1769–1832) and his
collaborators contributed to this new field a classic description
of the succession of rock formations that occurs in the Paris Basin.
Cuvier's work was subsequently taken note of by Oxbridge's
young geologists. Cuvier's table of strata, too, showed the actual
superposition and mineralogical composition of the rocks, but it
also added the fundamental observation that different rock units
contain different assemblages of fossils. The following decade
brought to the attention of the Geological Society of London the
highly detailed and empirical work of the English surveyor
William Smith (1769–1839), who had documented sequences of
strata in England, Wales, and Scotland and, like Cuvier, pointed
to the value of fossils in characterizing these strata. At this time,
however, little more existed than uncorrelated local or regional
rock profiles. Even Werner's table, in spite of its claims to global

applicability, was based on little more than the study of one region.

Several fellows of the Geological Society of London believed that an international standard stratigraphy was impossible. But they were proved wrong by Buckland. The end of the Napoleonic Wars stimulated international travel, and Buckland visited continental Europe a number of times. Cooperating with colleagues and students, in particular with the Anglican clergyman and geologist William Daniel Conybeare (1787–1857), Buckland produced a stratigraphic table that put together the various sections of the superposition of strata in the British Isles, making possible a comparison and correlation with rock successions in Werner's Saxony, Cuvier's Paris Basin, the Vienna Basin, the St. Petersburg region in the east, and parts of Italy in the south of Europe. Buckland collated the European data in an effort "to prove their identity with English formations by the evidence of actual sections; and to show that a constant and regular order of succession prevails in the alpine and transalpine districts, and generally over the Continent, and that this order is the same that exists in our own country."[14]

Thus the prime criteria for stratigraphic classification and correlation were superposition and lithology, a fact reflected in such formation names as "Coal Measures" and "Chalk." Buckland was well acquainted with Cuvier's and Smith's discoveries that rock units can be characterized by their fossil content, but these discoveries did not yet add up to an independent criterion of the relative ages of strata and thus of their place in the geologic column.

Through the 1820s a significant further development took place, namely, the intercontinental correlation of rock formation by an extension of the stratigraphy of Britain and the European mainland to other parts of the world. The person to establish the essential similarity of the European and American rock successions was Alexander von Humboldt (1769–1859). One of the fruits of his famous journey of exploration in the equatorial

Americas was a monograph almost instantly translated into English under the title *Geognostical Essay on the Superposition of Rocks, in Both Hemispheres* (1823). In this work he described "the most striking analogies in the position, composition, and the included organic remains of contemporary beds."[15] Humboldt—it should be added—although not a creationist as Cuvier and Buckland were, was not an evolutionist.[16]

Soon several magnificent, in some cases multicolored, "ideal" sections of the earth's crust were printed, one of them in the form of a foldout to Buckland's so-called Bridgewater Treatise on *Geology and Mineralogy Considered with Reference to Natural Theology* (1836), "intended to shew the order of deposition of the stratified rocks." Cuvier had observed not only that certain strata contain characteristic assemblages of fossils but also that the lower and thus older organic remains were taxonomically lower and that the younger assemblages included successively higher organisms. Buckland interpreted this "progressive development" of fossils as the effect of a slowly cooling globe, a conclusion he based on the "geotheory" of the French naturalist, the Comte de Buffon (1707–1788). In the course of a succession of geologic periods, each of them terminated by a worldwide catastrophe as postulated by Cuvier, the earth had become gradually more habitable to advanced animals and plants and ultimately to humans. During earlier and hotter times, reptiles had flourished, while a later and cooler world had allowed for the reign of mammals. The latest of all to appear on earth had been humankind, the highest of all.

Thus organisms past and present had been and were ideally adapted to their physical environments. In such adaptations Buckland saw compelling evidence of divine design. When in the geologic past a world catastrophe had wiped out a then-living creation, God had repopulated the world with new species, perfectly adapting them to the changed physical conditions of the surface of the earth. A tour de force of natural theology, Buckland's Bridgewater Treatise demonstrated the perfection of functional adaptations of fossils, such as extinct ammonites or giant sloths. More-over, the discontinuous nature of the fossil record was believed to disprove any notion of a transformation of species and instead to prove repeated revolutions of extinctions followed by divine, cre-ative interventions. The notion of fossil progress was, by and large, a conclusion drawn from the stratigraphic column and was not a belief used to construct that column.[17] In short—and to repeat—the theory of evolution had nothing whatsoever to do with the construction of the geologic column during the decades in which it acquired its largely definitive form.

The same applies to homology.[18] The person who most com-prehensively established homology, especially for the vertebrate classes of fishes, amphibians, reptiles, birds, and mammals, including humans, was the Anglican comparative anatomist Richard Owen (1804–1892), a protégé of Buckland's. Like his Oxonian patron, Owen drew on much previous work, mainly from continental Europe, but giving it a peculiarly British, natural-theological twist. The center of comparative anatomy in England was the Hunterian Museum of the Royal College of Sur-geons, where Owen worked and gained a reputation as Britain's leading biologist.

Owen began working systematically on problems of homo-logical morphology in 1841, as part of his curatorial task to arrange the osteological collection of the Hunterian Museum. Cataloging provided him with the materials for his exhaustive account of comparative osteology, presented in 1846 to the British Association for the Advancement of Science in the form of a major report and later published in book form under the title *On the Archetype and Homologies of the Vertebrate Skele-ton* (1848). Owen defined and illustrated a so-called vertebrate archetype—a blueprint or architectural plan—and formulated a precise meaning for the often synonymously used terms *homol-ogy* and *analogy*.[19]

The vertebrate archetype represented the generalized and sim-plified skeleton of all backboned animals. Owen demonstrated that the human skeleton can be traced—from top to bottom, throughout its extremities, and in its most complex and minute components—to the bony frames not just of other mammals but also of birds and reptiles, and even to the fishbones of the lowly salmon on our dinner plate. All vertebrates appeared connected, bone for bone, by invisible threads. With Owen's efforts, homology acquired a basis of systematic rather than incidental fact.

Throughout the 1830s and early 1840s Owen was an outspoken and well-known advocate of the doctrine of the creation of species and a supporter of Cuvier's criticism of transformist ideas. Owen's homological work opened a door for him to move away from special creation, however, and toward thinking about species as products of natural processes. By seeking the meaning of organic form not in the details of specific adaptations but in general archetypes, the argument from design was placed on a new, more abstract footing. God was no longer the supreme designer of Buckland's many functional adaptations but of Owen's architectural blueprints. The archetype—Owen was encouraged by his Anglican patrons to believe—was a Platonic idea, a plan of creation in the mind of the Creator.[20] Divine contrivance was to be recognized not so much in specific adaptations to the external conditions of individual species as in the common ground plan of living beings.

Thus the argument of design no longer required a belief in the special creation of species, and Owen cautiously began to formulate a theory of theistic evolution. In his *On the Nature of Limbs* (1849) he argued that species had come into existence by a preordained process of natural laws.[21] Still, Owen remained a staunch opponent of French ideas of organic transformation and later excoriated Darwin for attributing the origin of species to natural selection.

Darwin went one step further than Owen and brought the vertebrate archetype in God's mind down to earth by turning it into an ancestor of flesh and blood, transferring Owen's vast labors on vertebrate homologies to his own theory of evolution by means of natural selection.[22] In consequence, homology did not just indicate ancestry but, additionally, became the criterion of ancestry—circular reasoning. Nevertheless, by incorporating Owen's work, Darwin used a body of homological inferences that had not been based on the notion of descent. Owen's criterion of homology—that of relative position in the overall plan of a skeleton—provided Darwin with a straight, noncircular argument for evolution. As in the case of the geologic column, the formulation of homology and the vertebrate archetype took place without the slightest input from evolution. To the contrary, it gradually fostered speculations about a nonmiraculous, evolutionary origin of species.

I am not arguing that the geologic column is as purely factual in origin as many secular paleontologists and stratigraphers may like. History shows, too, that the geologic column and, for that matter, the vertebrate archetype were ideological *constructions* as well as factual *reconstructions,* shaped and—one might add—misshaped by the limitations and peculiarities of the time, as scientific developments tend to be. A variety of nonempirical, ideological influences were constitutive in these constructions, including nationalism, Eurocentrism, creationism, German idealism, Christian theism, social progressivism, and other "isms."[23] Creationists are right when they draw attention to such influences, but notably absent from the various "isms" that originally helped shape the geologic column and establish the homological relatedness of vertebrates was evolutionism.

MYTH 16

THAT EVOLUTION DESTROYED DARWIN'S FAITH IN CHRISTIANITY—UNTIL HE RECONVERTED ON HIS DEATHBED

James Moore

The story has circulated for decades. Charles Darwin, after a career of promoting evolution and naturalism, returned to the Christianity of his youth, renouncing on his deathbed the theory of evolution. The story appears to have been authored by a "Lady Hope" . . . many have researched the story, and all have concluded it is probably an "urban myth" . . . A Christian can only hope that the seeds planted earlier took root at the end, and that he did place his faith in Christ before he died.

   —John D. Morris, "Did Darwin Renounce Evolution on His Deathbed?" (2006)

There is no doubt that the integrity of this lady is so out-standingly clear that it would be equivalent to blasphemy to call into question her word. Lady Hope was a Christian of the finest evangelical tradition . . . One may confidently conclude that Lady Hope did visit Darwin shortly before his death and that during this visit she did witness Darwin's renewed faith in the Christian Gospel.

   —L. R. Croft, *The Life and Death of Charles Darwin* (1989)

In meditating on [the] absence of any reference by the family to Lady Hope's visits or Darwin's change of faith, I can only suggest that there may have been a quite deliberate agreement within the

family to say nothing whatsoever about what would have been a
late and very unwelcome turn of events.

—Malcolm Bowden, *True Science Agrees with the Bible*
(1998)

Every shopping mall has a well-stocked Lost and Found department. In the great mall of the history of science, the Lost and
Found is piled high with faiths. Of course, scientists don't mislay
or recover their religious beliefs as they do umbrellas, but one
may be excused for supposing so in the case of Charles Darwin.

For many years, scholars have tried to pinpoint Darwin's fall
from grace. A few say he was an agnostic or atheist by the time
he sailed on H.M.S. *Beagle* in 1831. Most date his apostasy from
the months after the voyage in 1836–1837, or to the period of his
marriage in 1839, or even as late as 1842, when he first wrote out
his theory of evolution. All these datings presume that, in accepting evolution, Darwin must have set aside the religious beliefs in
which he had been raised and educated. His own testimony is sel-
dom heeded: "I never gave up Christianity until I was forty years
of age."[1] Darwin turned forty in 1849, long after developing his
theory of evolution by natural selection.

While scholars have generally backed Darwin's early "loss of
faith," a pious throng are hopeful that he found it again later in
life. Many untutored evangelicals claim that in old age Darwin
saw the light and on his deathbed repented of evolution and returned to Christianity. In this tradition, Darwin's mislaid faith
is the hottest property in the history of science Lost and Found
department; his recovering it, a trophy of God's grace to be displayed.

Started in 1915, the story of Darwin's newfound faith swept
like wildfire through the evangelical press. Tracts such as "Darwin on His Deathbed," "Darwin's Last Hours," "Darwin 'The
Believer,'" and "Darwin Returned to the Bible" spread the blaze,
and over one hundred fresh outbreaks have been detected during

the next eighty years. In the twenty-first century, the story still crops up in creationist literature and even, amazingly, in the correspondence columns of the London *Times*.[2] Across fundamentalist America and to some extent in the nether regions of evangelicalism worldwide, a belief persists that Darwin abandoned his life's work and returned to the faith of his youth.

The "lost" and "found" traditions need not jar. Many fundamentalists espouse both, admitting the early loss of faith but convinced that the mature Darwin repented of his youthful impetuosity. But on the whole the traditions are at odds. The early-loser faction champions secular "scientific" beliefs and evolution; the late-finder side defends traditional "biblical" beliefs and creationism. Both sides, ironically, defer to Darwin's "greatness" by crediting his judgment in a realm about which he disclaimed any authority, but in the end, the traditions cancel out. Like the proverbial Kilkenny cats, they consume each other, leaving only their telling tales. Though considerably more and better evidence exists for Darwin's early loss of faith than for his recovering it late in life, both traditions are ill-founded and deserve to die.

What, then, do historians and biographers know about Darwin's faith? How did his religious beliefs develop? What sense is there in seeing his Christianity as "lost and found"? Charles Robert Darwin (1809–1882), the second son of a freethinking doctor and a devout Unitarian mother, was christened in the Church of England. As a boy, he attended chapel with his mother and was sent to the school run by the minister. He then sat under a future bishop at Shrewsbury School before studying medicine at Edinburgh University. Here he had his first zoology lessons from an out-and-out evolutionist dedicated to abolishing the church and bringing about radical social change. Dr. Robert Grant's religious influence was minimal. When Charles dropped out of medicine, his father prescribed a stint at Cambridge University to train for the Church of England. Charles went up in 1828 with scarcely a qualm.

Cambridge professors were untainted by radicalism. The reverends John Stevens Henslow (1796–1861) and Adam Sedgwick (1785–1873) agreed that species and society were kept stable by God's will. Darwin accepted their teaching and reaped the rewards. Down on his knees collecting beetles with Henslow's botanical parties, he took the professor as his role model—the parson-naturalist. After Darwin undertook a geological field trip with Sedgwick, Henslow offered him a place on H.M.S. *Beagle*, and in 1831 Darwin's path to a country parish was diverted via a voyage around the world.

For five years Darwin dreamt of living the parson's life. His religious beliefs and practices remained conventional and, like his professors, he found no science in Genesis. Charles Lyell's (1797–1875) *Principles of Geology* convinced him that the earth's crust had been laid down over countless ages according to natural laws. He theorized about islands and continents and began to see himself as a geologist. Meanwhile, his vision of life on earth was transformed by three events. Wandering for the first time in a lush Brazilian forest, he had something like a religious experience. "No one can stand unmoved in these solitudes, without feeling that there is more in man than the mere breath of his body," he confessed, even while sensing that humanity and nature were somehow one.[3] In Chile, he survived a terrifying earthquake. Nature's power awed him; even a cathedral was not spared. Most moving was his encounter with the aborigines of Tierra del Fuego. Could these wild, naked nomads have come from the same Hand that created civilized Cambridge dons?

Before the voyage ended, Darwin saw that living species—like races of people, plants, and animals—might have come into existence by descending from one another. So much, it seemed, could be explained if the diversity and distribution of God's creatures had come about through natural processes rather than miracles.

Back in London, Darwin's best friends believed in a God who ruled through natural laws. God must have created species by some progressive law and Darwin set out to discover it. In 1837 he began reading voraciously, making notes. Most Christians prided themselves in believing that humans were specially created, but for Darwin God's law of evolution sufficed. It was "more humble" to have apes in our family tree, to believe that savages and civilized people alike were "created from animals."[4] Yet he realized that espousing creation by evolution could be dangerous. If one species changes, if one instinct can be acquired, then the "whole fabric totters & falls."[5] The traditional fabric of Christian beliefs about nature, God, and humanity would have to be rethought.

Darwin decided to pitch his theory to traditionalists by emphasizing its superior theology. A world populated by natural law was "far grander" than one in which the Creator constantly interferes.[6] It now was obvious: nature "selects" the adapted organisms prized by parson-naturalists as proof of God's design. These organisms survive the struggle for life, laid down as a law of nature by Rev. Thomas Malthus (1766–1834), to pass on their advantages. "Natural selection," Darwin called it: through "death, famine, rapine, and the concealed war of nature," God's laws bring about "the highest good, which we can conceive, the creation of the higher animals."[7]

The church was now the last thing on his mind. With his father's support, Darwin had married his cousin Emma Wedgwood and in 1842 moved their growing family into an old parsonage in rural Kent. Emma was a sincere Christian like her aunt, Charles's mother, Unitarian by conviction, Anglican in practice. Her fears for his eternal destiny remained a sad undercurrent of their life together. When Charles collapsed after his father's death in 1848, events came to a head. A stay at a spa worked wonders, but Charles returned home to see his eldest daughter taken ill. When ten-year-old Annie died tragically over Easter 1851, he found no comfort in Emma's creed. After years of backsliding, Darwin finally broke with Christianity (though he continued to believe in God). His father's death had spiked the faith; Annie's clinched the point. Eternal punishment was immoral. He would speak out and be damned.

The *Origin of Species* (1859) did not mention the word *evolution,* but Darwin used *creation* and its cognates over one hundred times. Opposite the title was a quotation about studying God's works as well as his Word, and another by a reverend Cambridge professor about "general laws" as God's way of working. Darwin ended in a rhapsody about the "grandeur" of viewing nature's "most beautiful and most wonderful" diversity as the product of "powers . . . originally breathed into a few forms or into one."[8] This reference played to traditionalists, but the tone and the terminology—even the biblical "breathed"— were not insincere. From start to finish, the *Origin of Species* was a pious work: "one long argument" against miraculous creation but equally a theist's case for creation by law.

In his long-awaited *Descent of Man* (1871), Darwin portrayed humans as evolving physically by natural selection and then intellectually and morally through the inherited effects of habit, education, and religion. "With the more civilized races, the conviction of the existence of an all-seeing Deity has had a potent influence on the advance of morality," so much so that "the birth both of the species and of the individual are equally parts of that grand sequence of events, which our minds refuse to accept as the result of blind chance."[9]

Darwin spoke more personally in the autobiography written for his family between 1876 and 1881. Unwilling to give up Christianity, he had tried to "invent evidence" to confirm the Gospels, which prolonged his indecision. Just as his clerical career had died a slow "natural death," so his faith withered gradually. But there was no turning back once the deathblow struck. His dithering crystallized into a moral conviction so strict that he could not see how anyone—even Emma—"ought to wish Christianity to be true." For if it were, "the plain language" of the New Testament "seems to show that the men who do not believe, and this would include my Father, Brother and almost all my best friends, will be everlastingly punished. And this is a damnable doctrine." In later years, the residual theism evident in the *Origin of Species* became worn down, and now with "no assured and ever present belief in the existence of a personal God or of a future existence with retribution and reward," Darwin felt he must be "content to remain an Agnostic."[10]

Evangelicals, ever mindful of the prodigal son who "was lost, and is found," were not content. Darwin's agnostic confession was published in 1887, five years after his burial in Westminster Abbey. Yet within thirty years a story was circulating about his deathbed confession of faith. The anecdote first appeared in Boston in August 1915 in the Baptist family magazine *Watchman-Examiner*. Sponsored by the editor, who had heard it at an evangelical conference, the story was written out at his request by the author, known as Lady Hope. Born Elizabeth Cotton in 1842, she was a woman of note on the temperance circuit, a former associate of the evangelist D. L. Moody, and the widow of Admiral Sir James Hope of the Royal Navy. In Britain she had carried on bountifully, reading the Bible from door to door and writing treacly tracts and novels. After the admiral's death in 1881, she married an elderly millionaire but continued to trade off her secondhand title. Her largesse grew lavish, her lifestyle grand. She swapped the temperance wagon for a motor car and raced back and forth to evangelical meetings at unholy speeds. Creditors finally caught up with her; her bankruptcy made the London *Times*. After being freed from the liability of her debts, she retreated in 1912 to New York City, ostensibly "to overcome the grief of her husband's death." Less plausibly, after the *Watchman-Examiner* story came out, she claimed that her flight was "to avoid the persecution of the Darwins."[11]

Not that Americans minded. Most knew nothing of Lady Hope; those who did forgave. What counted was her story, which had the ring of truth, like Scripture. Here is Darwin, aged and bedridden, cradling an open Bible, his head haloed by an autumn sunset. Lady Hope attends him, basking in the "brightness and animation" of his face; nodding as he describes the "young man with unformed ideas" he once was, who wondered "all the time over everything"; smiling as he speaks of "the grandeur of this Book" and of "CHRIST JESUS! . . . and his salvation."[12] The imagery is familiar, irresistible, overwhelming. This is a deathbed drama—Darwin has been converted from evolution back to Christianity!

Stories like this vouched for themselves and were mildly addictive. Evangelicals, for whom faith's last best test was the deathbed, had long fascinated themselves with collections of "dying sayings" and "dying hours," "last days" and "last words." Lady Hope tapped this huge voyeurs' market. Had Darwin not been her subject, her story would still have sold. Shrewdly crafted, it reported neither a death scene nor repentance, but it aped such tales to perfection by playing up the drama and playing down the date, some six months before Darwin died. It was a brilliant counterfeit. Bankrupt abroad, Lady Hope sought spiritual credit in the United States, and she got it, pressed down, full and overflowing.

Although much of it was fictitious, the original story cannot be dismissed as pure invention. It contained startling elements of authenticity: the view from the window, a summer house in the garden, Darwin's gaudy dressing gown and his afternoon nap. What is more, Lady Hope clung to the story, privately supplying further convincing details until her death in 1922. She claimed to have been conducting "Gospel and Temperance" meetings in Darwin's village, while staying with a "lady" who lived "very near" his house, and she knew the "large gate" that opened onto its "carriage drive." Darwin himself asked her to call at "three P.M.," his siesta hour, and she found him lying on a "sofa" beside a "fine bay window" in "a large room with a high ceiling" just off the upstairs "landing."[13] Why, if sensational self-aggrandizement was her aim, did Lady Hope not incorporate these details into the original story?

The upshot (after much research) is that either Lady Hope visited Darwin at home as she claimed—no matter what they discussed—or she was able to winkle enough scraps of intimate information from his loyal domestic staff to render a preposterous story plausible. Neither alternative is likely to appeal to the proponents of either tradition—Lost or Found—about Darwin's faith, or to those who hope to have it both ways. But since the traditions, like Kilkenny cats, tend to cancel each other out, an alternative tack may help us get beyond the impasse in thinking about Darwin's Christianity.

Just suppose that one's religious faith is judged by actions more than words, by deeds more than creeds; that being a Christian is as much or more about doing than believing. Suppose, in short, that the Epistle of James is right, "faith without works is dead," and that the pious doctrinaire may be challenged, "Shew me thy faith without thy works, and I will shew thee my faith by my works." Grant this (as many Christians through the ages would) and it instantly renders both Lost and Found traditions irrelevant.

The English lay no one lightly in Westminster Abbey, their national shrine, much less the mortal remains of those who affront the monarchy, the established church, or Christianity. Darwin's contemporaries George Eliot (who lived in sin with her partner George Lewes) and Herbert Spencer (who hated all establishments) were excluded, so how did a self-confessed agnostic get in?

Darwin cut an establishment figure. Christened in the church, he studied for Anglican orders at Cambridge and after the *Beagle* voyage he retired to an old parsonage. His infants were christened and the family attended the parish church. He himself fell away, but he gave generously toward church repairs and sent his boys to be tutored by clergymen. Local priests always had his support; the Reverend John Innes became a lifelong friend. In 1850 they started a benefit society for the parish laborers, with Darwin as guardian. Innes later made him treasurer of the local charities and, with a testimonial from him in 1857, Darwin became a county magistrate, swearing on the Bible to keep the

Queen's peace. His grandmother had died an alcoholic and he shared Lady Hope's concern for drunkenness. He turned a parish schoolroom over to her co-worker, James Fegan, for gospel meetings, altering the family's dinner hour so the servants could attend, and Emma visited his mother at home.[14] Through the years Darwin quietly supported liberal churchmen and even sent annual donations to the Church Missionary Society for the Anglican mission in Tierra del Fuego. He never published a word directly against Christianity or belief in God. The *Origin of Species* was the last major work in the history of science in which theology played an active role, and thus the Church of England had no fear "lest the sacred pavement of the Abbey should cover a secret enemy." Darwin's teaching was seen as consistent "with strong religious faith and hope" and his burial at Westminster as a visible sign of "the reconciliation of Faith and Science."[15]

If Christian is what Christian does, Darwin kept the faith of Victorian England.

THAT HUXLEY DEFEATED WILBERFORCE

IN THEIR DEBATE OVER EVOLUTION AND RELIGION

David N. Livingstone

[T]he first major battle in a long war.

—John H. Lienhard, "Soapy Sam and Huxley" (1998)

The zoology and botany section of the British Association for the Advancement of Science met on Saturday, June 30, 1860, in the library of Oxford's new university museum. Charles Darwin's *Origin of Species,* published the previous November, was the subject of a discussion chaired by John Stevens Henslow (1796–1861), Cambridge University's professor of botany. What supposedly transpired that afternoon has often been told and nowhere more vividly than in the following account by William Irvine, professor of English at Stanford University. As befits a literary expert and amusing raconteur, Irvine's prose is colorful, indeed racy, and altogether memorable, as can be gleaned from the following excerpts:

> Bishop Wilberforce, widely known as "Soapy Sam," was one of those men whose moral and intellectual fibres have been permanently loosened by the early success and applause of a distinguished undergraduate career. He had thereafter taken to succeeding at easier and easier tasks, and was now, at fifty-four, a bluff, shallow, good-humoured opportunist and a formidable speaker before an undiscriminating crowd . . . Finally, overcome by success, he turned with mock politeness to Huxley and "begged to know, was it

through his grandfather or his grandmother that he claimed his descent from a monkey?" This was fatal. He had opened an avenue to his own vacuity. Huxley slapped his knee and astonished the grave scientist next to him by softly exclaiming, "The Lord hath delivered him into mine hands." The Bishop sat down amid a roar of applause and a sea of fluttering white handkerchiefs. Now there were calls for Huxley . . . He touched on the Bishop's obvious ignorance of the sciences involved; explained, clearly and briefly, Darwin's leading ideas; and then, in tones even more grave and quiet, said that he would not be ashamed to have a monkey for his ancestor; but he would be "ashamed to be connected with a man who used great gifts to obscure the truth." The sensation was immense. A hostile audience accorded him nearly as much applause as the Bishop had received. One lady, employing an idiom now lost, expressed her sense of intellectual crisis by fainting. The Bishop had suffered a sudden and involuntary martyrdom, perishing in the diverted avalanches of his own blunt ridicule. Huxley had committed forensic murder with a wonderful artistic simplicity, grinding orthodoxy between the facts and the supreme Victorian value of truth-telling. At length Joseph Hooker rose and botanized briefly on the grave of the Bishop's scientific reputation.[1]

Often compressed, sometimes embellished, usually with minor variations, this tale has been repeated for popular audiences for nearly a century and a half. It was parodied in Charles Kingsley's *Water Babies* in 1863, performed on BBC television's serialized drama *The Voyage of Charles Darwin* in 1987, and rerun at the 1988 British Association (BA) meeting in Oxford by Bishop Richard Harries and the geologist Beverly Halsted. Not surprisingly it featured, albeit rather tersely, in Andrew Dickson White's 1896 *A History of the Warfare of Science with Theology*. According to White, Wilberforce congratulated himself at the meeting

that he was not descended from a monkey. The reply came from Huxley, who said in substance: "If I had to choose, I would prefer to be a descendant of a humble monkey rather than a man who employs his knowledge and eloquence in misrepresenting those who are wearing

out their lives in the search for truth." This shot reverberated through
England, and indeed through other countries.[2]

Sheridan Gilley's observation on the depiction of the scene in
"the vulgar mythology of the television screen" evidently has
wide application: in these scenarios "Huxley and Wilberforce
are not so much personalities as the warring embodiments of rival moralities, Huxley, the archangel Michael of enlightenment,
knowledge, and the disinterested pursuit of truth; Wilberforce,
the dark defender of the failing forces of authority, bigotry and
superstition."[3]

Despite the correctives that have been issued over the years
by historians seeking to expose its mythological character, the
story continues to have symbolic currency within the scientific
world. As recently as 2004, when the physicist M. M. Woolfson
presented a lively—and telling—critique of the problems of
conformism and peer pressure in scientific research, he paused
to recount "the debate that took place . . . between Thomas
Henry Huxley (1825–1895), a Darwin supporter, and Bishop
Samuel Wilberforce (1805–1873) of Oxford, a vehement opponent of Darwin's ideas." "Huxley's arguments" he went on,
"were the more cogent and he convinced many unbiased members of the audience, but the debate ended with considerably
more than one-half of the audience supporting the bishop."[4] As
a rhetorical tactic to invite readers to "consider which is the
better—conforming or thinking," Woolfson's history lesson
called forth a pointed response from the historian Frank James,
who insisted that it was just "very hard to tell what did happen" that afternoon, "not only because of the paucity of contemporary sources, but also because these may not have been
accurate." Besides, there was no vote taken from which it could
be determined what proportion of the audience supported what
position. The myth, he pointed out, was "created 20 years later
and, despite the best efforts of historians, it is still trotted out
uncritically to this day."[5]

It is indeed impossible to know exactly what went on in the Oxford Natural History Museum that summer day, and much of the story as it has come down to us is a fabrication.[6] It was stitched together many years after the event and is largely the product of the big Victorian biographies of the likes of Darwin, Huxley, and William Hooker (1785–1865) that were published decades later. The *Life and Letters of Thomas Henry Huxley,* which was assembled by his son Leonard in 1900, perhaps most of all crystallized the standard perception; to him the encounter was "an open clash between Science and the Church."[7] Moreover, while Isabel Sidgwick, who was present at the event, recalled thirty years later Huxley's comment about not being ashamed to have an ape for an ancestor, the only extensive contemporary report, which appeared a couple of weeks after the meeting in the *Athenaeum,* contained no reference to grandfathers, grandmothers, or ape relations.[8] By the same token, the fact that *The Press,* in its brief commentary on July 7, 1860, reported Wilberforce asking Huxley "whether he would prefer to have a monkey for his grandfather or his grandmother" has encouraged some to think that the *Athenaeum*'s account could well have been sanitized to preserve the gentlemanly reputation of the BA.[9]

Taken overall, reports at the time are remarkably thin on the ground and those that do exist contradict one another. For a start, each side felt sure that victory was theirs. Huxley, Darwin's "bulldog," was confident that he had carried the day and believed he "was the most popular man in Oxford for a full four & twenty hours afterwards." For his part, Wilberforce was sure that he had thoroughly roasted Huxley. By contrast, the botanist Joseph Hooker congratulated *himself* on securing the conquest. "I smashed him amid rounds of applause. I hit him in the wind at the first shot in ten words taken from his own ugly mouth," he wrote to Darwin a few days after the meeting. "Sam was shut up—had not one word to say in reply, and the meeting was dissolved forthwith."[10] As for any electric shock surging through the audience as a result of Huxley's sparky oratory, it is well to remember that his voice was thought to be too weak to hold a large company of listeners. And Hooker did not hesitate to mention it. Whatever he himself felt about the ugliness of the bishop's oration and the worthiness of the bulldog's answer, he nonetheless felt compelled to report to Darwin that Huxley "could not throw his voice over so large an assembly nor command the audience . . . [H]e did not allude to Sam's weak points nor put the matter in a form or way that carried the audience."[11]

In the midst of these swirling impressions and irresolutions, one thing is certain: the staging of the confrontation as the *pièce de résistance* in the epic war between science and religion is altogether misguided. For one thing, some clerical members of the audience gave their support to Huxley's side of the debate even if Henry Baker Tristram found Wilberforce sufficiently convincing to persuade him to give up Darwinism.[12] Frederick Temple, later archbishop of Canterbury, who preached on the relations between science and religion in the University Church during the BA meeting, welcomed the reign of natural law and thereby allowed space for the development of an evolutionary reading of nature, a view he reiterated later in his Bampton Lectures. Besides, one of Darwin's earliest supporters was the novelist and clergyman Charles Kingsley. At the same time there remained considerable scientific opposition to Darwin's proposals. The fact that Darwin himself believed he could discern the hand of the anatomist Richard Owen (1804–1892) behind Wilberforce's stinging review only serves to confirm that the feud was emphatically not simply a struggle between science and religion. It was as much as anything else a clash between different scientific visions. Indeed, the scene for the Saturday scuffle had been set on the previous Thursday, when Huxley had crossed swords with Owen over the question of the similarity between ape and human brains. On the Saturday afternoon, too, Lionel Beale, medical practitioner and professor at King's College London, pointed to several difficulties in Darwin's theory. Besides, throughout his assessment for the *Quarterly Review,* Wilberforce persistently called on the testimony of scientific practitioners, not scriptural authority.[13] As he insisted, "we have objected to the view with which we are dealing solely on scientific grounds . . . We have no sympathy with those who object to any facts or alleged facts in nature . . . because they believe them to contradict what it appears to them is taught in Revelation."[14]

Darwin's own assessment of the bishop's review of *Origin of Species,* which came out in the *Quarterly* just a day or two after the debate—a review that had been written some five weeks before the Oxford meeting—is also telling; he conceded that it was "uncommonly clever," picking out "all the most conjectural parts" and highlighting real "difficulties."[15] Indeed, Wilberforce's allergy to scientific speculation as a gross violation of Baconian induction found support from the aging geologist Adam Sedgwick (1785–1873), who had already written to Darwin lamenting that he had "*deserted* . . . the true method of induction, and started machinery as wild . . . as Bishop Wilkins's locomotive that was to sail with us to the moon."[16] For all these reasons it is not surprising that Huxley's recent biographer, Adrian Desmond, concludes that any idea of a clear victor is mistaken and that the *Athenaeum*'s judgment that Wilberforce and Huxley were well matched is not far off the mark.[17]

That there was some kind of altercation that June afternoon is not in doubt, but various contextual factors cast the meaning of the incident into significant relief. The first complication rotates around the struggle to professionalize science during the Victorian era. Scientific endeavor had been in a weak position in Oxford during the first half of the century, and various factions had been pressing for the strengthening of the natural sciences over the previous twenty years or so. The 1860 BA meeting in the new museum marked the symbolic public dedication of Oxford's new cathedral of science, and it provided the perfect emblematic location to drive home the point that science was at last throwing off the shackles of traditional authority. The emergence of a younger, professional class of scientists seeking to wrest cultural power from the older, amateur clerical naturalists (Wilberforce was fifty-four years old, Huxley thirty-five) meant that men like Wilberforce—whatever their accomplishments—were no longer welcome among the new breed of specialists.[18] The need to marginalize the older generation in the emerging knowledge economy manifested itself in the dramatic decline in the number of Anglican clergy who held presidency of the BA during the later decades of the nineteenth century. As John Hedley Brooke has pointed out, forty-one had presided over the association in its first three and a half decades; in the period from 1866 to the turn of the century, the number dropped to three. The feud, to put it another way, was between different styles—and different cohorts—of scientific practice.

There were, too, factional struggles within the Anglican communion that cast light on the BA machinations. Wilberforce, by no means popular among his fellow churchmen by 1860, especially in Oxford, was increasingly distressed by liberalizing trends within the Church of England. He was chagrined to find fellow Anglicans authoring *Essays and Reviews* (1860), a tract for the times that gave support to the new German higher criticism; to him this represented nothing less than a complete capitulation to infidel continental metaphysics. That prominent clergy such as Temple, Rowland Williams, and Henry Bristow Wilson should join forces with scholars like Baden Powell, Mark Pattison, and Benjamin Jowett to promulgate such heresy prompted Wilberforce to issue a stinging refutation in the *Quarterly Review*. For Wilberforce, opposition to Darwinian naturalism was all of a piece with his alarm over modernizing impulses within theology. Nor was this the peculiar passion of the bishop of Oxford. A letter to *The Times* signed by the archbishop of Canterbury (the evangelical John Bird Sumner) and twenty-five Church of England bishops threatened the theologians involved with ecclesiastical censure—a move that prompted Darwin to quip, using his favorite proverb, that "a bench of bishops is the devil's flower garden."[19] Along with John Lubbock (1803–1865), Charles Lyell (1797–1875), and other members of the new scientific elite, Darwin signed a counter-letter supporting the volume. Darwinian evolutionists and liberal churchmen were joining together in a common cultural project.

Whether or not decorum was breached during the row also has a bearing on how the event was seen. Matters of etiquette and good taste were certainly in the minds of some who reflected on the occasion. Frederic William Farrar (1831–1903), canon of Westminster Abbey, recalled that what the bishop said was neither vulgar nor insolent, but flippant, particularly when he seemed to degrade the fair sex by pondering whether anyone— whatever they thought about their *grandfather*—would be willing to trace their descent from an ape through their *grandmother*. In Farrar's opinion, everyone recognized that the bishop "had forgotten to behave like a gentleman" and that Huxley "had got a victory in the respect of *manners* and *good breeding*."[20] James Young Simpson concurred, observing in his *Landmarks in the Struggle between Science and Religion* (1925) that in the exchange "the honours . . . in common courtesy" lay with the bulldog, not with the bishop.[21] Whatever his precise words, Wilberforce "used the age-old formula of delivering an insult disguised as a friendly jest" and, as Janet Browne notes, the "gibe was understood by every member of the audience."[22] And yet, while later writers placed Huxley on the side of good breeding, at the time both the *Athenaeum* and *Jackson's Oxford Journal* thought him discourteous. The boundaries of civility and politesse shifted over the decades. As Paul White puts it, Huxley's frankness "still seemed unruly and discreditable" in 1860, while Owen, who had "once seemed honest and polite, appeared disreputable and ill-mannered" by the end of the century.[23]

All of these comments point to the significance of the dispute as a *rhetorical event* and of the intimate connections between location and locution in scientific communication.[24] Indeed, the Oxford meeting was not the only time when speechifying at the BA fell foul of local propriety. In Belfast fourteen years later, both Huxley and John Tyndall (1820–1893) were excoriated for their display of tastelessness. William MacIlwaine, speaking to the Belfast Naturalists' Field Club in November 1874, made it clear that he thought Tyndall's famous address—promising to "wrest from theology, the entire domain of cosmological theory"—"reckless" and "a violation of the rules of good taste," while the local *Almanack* declared that both Tyndall and Huxley had "exhibited very bad taste" at the previous summer's meeting.[25] What can, and cannot, be said, at least with impunity, in public arenas shapes the way in which events are perceived. Whether by accident or design, pushing the boundaries of rhetorical propriety has the effect of attracting attention and securing a hearing, while at the same time concealing substantive issues under a cloud of surface noise. Juicy oratory and racy rhetoric at the Oxford BA allowed fading memories to make myths.

In narrating and reinterpreting a tale so pregnant with symbolism, it should not be inappropriate to end with what might well be taken as a symbolic—and indeed ironic—silence. On that fateful afternoon in June 1860, John William Draper suffered the misfortune of presenting a paper that, reportedly, was by turns boring and irritating. It was an attempt to use Darwinian vocabulary to explain what he called the intellectual development of Europe. As conventional wisdom now has it, his assertion "Let this point A be man and that point B be the mawnkey" was treated to the indignity of taunting cries of "mawnkey" from the assembled audience just a few minutes before Wilberforce took center stage. Hooker described Draper's efforts as the meanderings of a "Yankee donkey" and only waited around because he knew that Wilberforce was soon to take the floor. Nevertheless, a decade and a half later, when Draper published his epic *History of the Conflict between Religion and Science,* that iconic clash between Wilberforce and Huxley as representatives of the forces of religion and science is conspicuous—but only by its absence.[26]

THAT DARWIN DESTROYED NATURAL THEOLOGY

Jon H. Roberts

Explaining the perfection of adaptation by materialistic forces
(selection) removed God, so to speak, from his creation. It
eliminated the principal arguments of natural theology, and it
has been rightly said that natural theology as a viable concept
died on November 24, 1859.

—Ernst Mayr, *The Growth of Biological Thought* (1982)

The mortal damage to natural theology came from Darwin's
hypothesis of evolution by the mechanism of species . . . The
evidence of the operation of chance, brutality, suffering and
extinction changed the view of the universe for many Victorians
and destroyed the reverent attitude and sacred image of the
world necessary for natural theology.

—T. M. Heyck, *The Transformation of Intellectual Life
in Victorian England* (1982)

[A]lthough atheism might have been *logically* tenable before
Darwin, Darwin made it possible to be an intellectually fulfilled
atheist.

—Richard Dawkins, *The Blind Watchmaker: Why the
Evidence of Evolution Reveals a Universe without Design*
(1986)

During the century prior to the publication of Charles Darwin's
Origin of Species in 1859, natural theology, broadly defined as
the effort to establish God's existence and attributes through the

use of reason, played an important role in Christian discourse, especially in the English-speaking world. Scientists, theologians, and clergy alike employed the insights of natural theology to advance a variety of projects, including the defense of theism against the claims of unbelievers, the demonstration of the value of science, the establishment of common ground among Christians of different ecclesiastical traditions, and the promotion of piety evoked by the wondrous benevolence and wisdom embodied in the creation.[1]

Historically, *natural theology* has served as an umbrella term for a wide variety of arguments. These have ranged from purely rationalistic "ontological" arguments claiming that the very meaning of the concept of God as a perfect Being implies the existence of such a Being, on the grounds that existence is an inherent element of perfection, to "cosmological" arguments that argue from the contingency of the universe to the existence of a "necessary Being." During the first three-quarters of the nineteenth century, the particular argument that dominated discussions of natural theology in Great Britain and the United States was the argument from design. There were two major versions of this argument, and both derived much of their broad appeal from the fact that they drew primarily on characteristics from the organic world. The first and probably most popular rendition emphasized the usefulness of virtually all of the characteristics of plants and animals in helping to adapt those organisms to the environments in which they found themselves. For proponents of this "utilitarian" form of the argument from design, each instance of adaptation seemed to constitute additional testimony to divine wisdom and goodness. The other version of the argument focused on the pervasive existence of intelligible patterns within the organic world. Partisans of this "idealist" form of the argument from design held that the history of life could best be understood as the gradual material realization of a premeditated and integrated plan that had been formulated by a benevolent and rational Deity. Some natural theologians used both of these formulations of the argument from design in their defense of theism.[2]

By providing a naturalistic explanation of both adaptation and the "unity of type" and other harmonious patterns invoked by proponents of the idealist form of the argument from design, Charles Darwin's (1809–1882) theory of descent by means of natural selection challenged both the concept of designed adaptation and the notion of a premeditated plan. Recognizing this, a number of Darwin's supporters and critics alike were quick to suggest that Darwin's work fatally undermined the argument from design. "Darwin's bulldog," Thomas Henry Huxley (1825–1895), thus asserted in 1864 that "Teleology, as commonly understood [to mean design and purpose], had received its deathblow at Mr. Darwin's hands."[3] On the other end of the theological spectrum, a concerned Daniel R. Goodwin, provost of the University of Pennsylvania, dolefully predicted that in destroying "the *marks*, the *proofs* of design, and consequently the *evidence* of an intelligent controlling cause," Darwin's theory would "surely breed atheism and pantheism" in its supporters.[4]

Apparently taking such claims at face value, a number of students of the Darwinian controversy have concluded that Darwin's work dealt a death-blow to the enterprise of natural theology itself. That conclusion is unwarranted. Indeed, even a cursory examination of the history of Christian thought in the Anglo-American world since 1859 is sufficient to indicate that natural theology remained an ongoing, sometimes even thriving, enterprise. Leaving aside the question of the validity of the arguments that have been advanced in the name of natural theology and the uses to which those arguments have been put, I shall here briefly trace the persistence of natural theology since the appearance of the *Origin of Species*. Because it seems reasonable to believe that Darwin's theory had little impact on the fortunes of natural theology in the work of those who rejected that theory, I have chosen to focus my discussion on the views of the numerous theists who have accepted the transmutation hypothesis.

During the last quarter of the nineteenth century a sizable number of proponents of the theory of organic evolution—the theory that most scientists and theologians alike at that time equated with "Darwinism"—put forward a variety of arguments that clearly fell within the purview of natural theology. Some of those arguments continued to display the fascination that earlier nineteenth-century natural theologians had exhibited with the design of the organic world. Two such arguments proved to be especially popular. One argument, which was first articulated by the Harvard botanist Asa Gray (1810–1888) in 1860, focused on the issue of variability. Proponents of this argument held that while Darwin's theory of natural selection might well be able to explain the survival of the fittest, it could not explain the origin of the variations on which natural selection operated. Thus they insisted that evolution could not be a helter-skelter "method of trial and error in all directions" but must involve the oversight of "a directing mind."[5] Darwin found this argument sufficiently vexing to the spirit of his theory to devote space refuting it in *The Variation of Animals and Plants under Domestication* (1868).[6]

The other popular argument put forward by partisans of natural theology who drew on insights from evolution was predicated on a careful parsing of the concept of natural selection. Noting that this concept was nothing more than, as Gray put it, a "generalized expression for the processes and the results of the whole interplay of living things on the earth with their inorganic surroundings and with each other," they held that it was inconceivable that "blind, automatic nature, without a controlling will, could have blundered through these myriads of transformations with any show of success or regularity."[7] Natural theologians insisted, rather, that the "union and conspiracy of forces involved in Evolution"—the "useful collocations" and "fit arrangements" of events and processes that had culminated in the emergence of an "infinite variety of organic adaptations" during the course of the history of life—could be explained adequately only if they were seen as the product of a divine "coördinating power."[8] The fact that the evolutionary process appeared to be "progressive" in that it revealed the appearance of ever "more highly organized species" seemed to constitute additional grounds for suggesting that natural selection "not only does not conflict with the argument of design, but affords new illustration of it."[9] A similar conviction prompted James McCosh, who served as the president of Princeton College, to assert that "supernatural design produces natural selection."[10] British apologists concurred. The Scottish Free Church Presbyterian James Iverach, for example, maintained that natural selection was "but another way of indicating design."[11]

Despite the popularity of the claim that Darwin's theory of the history of life made sense only if it were placed within a broader theistic framework, many defenders of belief in God in the late nineteenth century chose to move beyond the organic world in framing their arguments. It may well be that the most important role that Darwinism played in shaping the fortunes of natural theology lay in convincing many theists to broaden the scope of the design argument from living things to the intelligibility of the natural world as a whole. Such theists pointed out that the existence of a universe that was sufficiently pervaded by pattern and order to be describable in terms of natural laws was not a logical necessity. Nor was it an outcome that could reasonably be expected from interactions of particles and forces acting at random.[12] Thus, they maintained that "order is invariably conjoined with intelligence" and that "that which is intelligible has intelligence in it."[13] From this perspective, the very regularity with which gravitation and other natural processes operated provided eloquent testimony to the pervasiveness of divine design. It therefore seemed reasonable to conclude, as one Unitarian clergyman put it, that "the more law, the more God, the more mystery, wonder, awe, and trust."[14]

By the middle of the 1880s, many Anglo-American evolutionists had managed to convince themselves that Darwinism "does not touch the great truths of natural theology nor can it touch them, except as it gives us new materials with which to prove them."[15] Lewis E. Hicks, a professor of theology at Denison College who wrote an extensive analysis of the argument from design, declared in 1883 that "there is no longer any room for doubt upon the general question whether the acceptance of evolution is destructive to all design-arguments. Theologians are practically a unit in the feeling that a belief in evolution leaves theism intact."[16] And theologians were not alone. Although the canons of professionalism led an ever-growing number of scientists to eschew God-talk in their professional publications, this did not prevent them from espousing the legitimacy of natural theology in works directed at the general reader. For example, the University of California natural historian Joseph Le Conte (1823–1901), perhaps the most influential theistic evolutionist in America, described science itself as "a rational system of natural theology" in that it pointed beyond itself to a divine Mind that served as the "energy" that was immanent throughout creation.[17] In addition, as Bernard Lightman has shown, some of the best-selling popular science books published during the late nineteenth century made natural theology a subtle but important subtext.[18]

During the period after 1900 natural theology enjoyed continued vigor within a number of important communities of discourse in Great Britain and North America. It remained vibrant among Anglo-American philosophers of religion and neo-Scholastic theologians.[19] A number of scientists also published statements lending credence to the notion that the natural world attested to the existence of a divine Creator and Designer. During the early twentieth century, a variety of British religious writers, who supported a "new" natural theology, drew on the work of biologists and psychologists who expressed impatience with mechanistic materialism.[20] A bit later, some well-known physicists lent their support to the cause. During the early thirties, Sir James Jeans

(1877–1946), a fellow of the Royal Society who taught applied mathematics at Princeton and Cambridge before becoming an independent scholar, inferred from his conviction that the cosmos appeared to be "more like a great thought than like a great machine," that it had been designed by a "universal mind," which he identified as "the Great Architect of the Universe."[21] Liberal clergy also sometimes appealed to data drawn from the natural world in attempting to convince their parishioners of the legitimacy of the theistic world view. The prominent New York City clergyman Harry Emerson Fosdick, for example, justified the credibility of theism on the grounds that it made far more sense than the belief that the "whole creative process" could be described in terms of "the fortuitous interactions of a few chemical elements."[22]

Notwithstanding the persistence of expressions of allegiance to natural theology among a variety of thinkers, interest in that enterprise waned somewhat during the first two-thirds of the twentieth century. This was largely because many theistic evolutionists in North America and Great Britain alike during that period had come to believe that the conclusions of natural theology were ill-suited to foster devotion to the God of Christianity. Drawing a good deal of inspiration from German theologians such as Friedrich Schleiermacher and Albrecht Ritschl, those Protestants concluded that the foundation of the Christian world view resided not in inferences drawn from the natural world but in the realms of feelings and values. During the 1920s and beyond, the growing influence of neo-orthodoxy and other theologies that drew inspiration from the perspective of Karl Barth prompted many Anglo-American theists to dismiss natural theology as an inadequate and even misleading approach to the God revealed by Jesus through the Word of God.[23]

Although efforts to establish God's existence and attributes through arguments using data derived from the natural world remain out of favor with many proponents of the Judeo-Christian world view, the last forty years have witnessed a resurgence of efforts to demonstrate the legitimacy and importance of natural theology. The theologian Nancey Murphy gave voice to the rationale underlying these efforts in 1990, when she asserted that in mounting an "effective apologetic strategy," it was essential to exploit the resources of natural theology and revealed theology alike.[24] Although Murphy is certainly not alone among theologians who have expressed interest in natural theology, many of the well-known exponents of arguments for theism drawn from the natural world in recent years have been scientists (albeit often ones who also possess credentials in theology). In contrast to the many Anglo-American theologians who have focused most of their attention in recent years on the nature of the divine-human encounter, those scientists have emphasized that, in the words of the British physicist-theologian John Polkinghorne, "there is more to God than his dealings with men."[25] Although they have little sympathy with the claim of partisans of the "Intelligent Design" movement that "design can be rigorously reformulated as a scientific theory," they have reasoned that "if God is the Creator of the world he has surely not left it wholly without marks of his character, however veiled."[26] Polkinghorne, for example, has insisted that the intelligibility of the universe, coupled with the fact that its precise, "tightly knit" structure permits the emergence of life against all reasonable probability, provides strong grounds for affirming the existence of a divine Mind "worthy of worship and the ground of hope."[27] Similarly, the late biochemist and Anglican priest Arthur Peacocke suggested that "from the existence of the kind of universe we actually have, considered in the light of the natural sciences," it is appropriate to "infer the existence of a creator God as the best explanation of all-that-is."[28]

During the period since the publication of the *Origin of Species,* the fortunes of natural theology have waxed and waned within Anglo-American culture. Darwin's theory was clearly instrumental in convincing natural theologians that they would be well advised to alter the kind of arguments that they employed in drawing inferences from nature to nature's God. It seems to have played a negligible role, however, in accounting for the shifting status of natural theology itself.

THAT DARWIN AND HAECKEL WERE COMPLICIT IN NAZI BIOLOGY

Robert J. Richards

[Haeckel's] evolutionary racism; his call to the German people for racial purity and unflinching devotion to a "just" state; his belief that harsh, inexorable laws of evolution ruled human civilization and nature alike, conferring upon favored races the right to dominate others; the irrational mysticism that had always stood in strange communion with his grave words about objective science—all contributed to the rise of Nazism.

—Stephen Jay Gould, *Ontogeny and Phylogeny* (1977)

No matter how crooked the road was from Darwin to Hitler, clearly Darwinism and eugenics smoothed the path for Nazi ideology, especially for the Nazi stress on expansion, war, racial struggle, and racial extermination.

—Richard Weikart, *From Darwin to Hitler* (2004)

In 1971, Daniel Gasman saw published his *Scientific Origins of National Socialism: Social Darwinism in Ernst Haeckel and the German Monist League,* the dissertation he had produced at the University of Chicago two years before. That book argued that Ernst Haeckel (1834–1919), the great champion of Darwinism in Germany, had special responsibility for contributing to Nazi extermination biology. Gasman stacked up the evidence: that Haeckel's Darwinian monism (which held that no metaphysical distinction separated man from animals) was racist; that he was a virulent anti-Semite; and that leading Nazis had adopted his monistic conceptions and racial views. Quite uncritically, scores of historians have accepted Gasman's claim, the most prominent of whom, at least among historians of biology, has been Stephen Jay Gould.

In his book *Ontogeny and Phylogeny* (1977), Gould investigated the consequences of Haeckel's "biogenetic law," the principle that the embryo of an advanced creature recapitulates the same morphological stages that the phylum went through in its evolutionary descent. According to Haeckel's law, a human embryo, for instance, begins life as something like a one-celled creature, then advances through the forms of an invertebrate, a fish, an ape, and finally a particular human being. Gould argued that the principle of recapitulation sustained an unwarranted progressivist interpretation of evolutionary theory and had racist implications. He urged that Charles Darwin (1809–1882) had refrained from adopting the principle, though acknowledged that many biologists had subsequently accepted it as part of the Darwinian heritage. The law, in Gould's estimation, was not Haeckel's most enduring legacy, however. Rather, "as Gasman argues, Haeckel's greatest influence was, ultimately, in another tragic direction—national socialism."[1]

Gasman's thesis has been used by religious fundamentalists as a crude lever by which to pry Darwinian theory away from public approbation. Put "Haeckel" and "Nazis" into any web search engine, and you will get thousands of hits, mostly from creationist and intelligent design websites that set alight Haeckel's Darwinism in an electronic auto-da-fé.

Most historians, save for Richard Weikart (quoted above), have refused to indict Darwin for complicity in the crimes of the Nazis. Gasman, Gould, and many other scholars have striven to distinguish Darwin's conceptions from those of Haeckel. In the nineteenth century, one individual of singular authority did not, however, detect any differences between the doctrines of the two biologists—namely, the English master himself. Early in their acquaintance, Darwin wrote Haeckel to say that "I am delighted that so distinguished a naturalist should confirm & expound my views; and I can clearly see that you are one of the few who clearly understands Natural Selection."[2] Their initial correspondence led to an enduring friendship, with Haeckel visiting Darwin several times at his home in the village of Downe. In *The Descent of Man,* Darwin affirmed their common understanding of evolutionary theory: "Almost all the conclusions at which I have arrived I find confirmed by this naturalist [Haeckel], whose knowledge on many points is much fuller than mine."[3] Though their emphases certainly differed, Haeckel and Darwin essentially agreed on the technical issues of evolutionary theory.[4]

If the indictment of complicity with the Nazis stands against Haeckel, should it then be extended to include Darwin and evolutionary theory more generally? Did Haeckel simply pack Darwin's evolutionary materialism and racism into his sidecar and deliver their toxic message to Berchtesgaden as Weikart has recently maintained?[5] Let me answer these questions by considering their subsidiary parts: Was Darwinian theory progressivist, holding some species to be "higher" than others? Was it racist, depicting some groups of human beings to be more advanced than others? Was it specifically anti-Semitic, casting Jews into a degraded class of human beings? Did Darwinian theory rupture the humanitarian tradition in ethics, thus facilitating a depraved Nazi morality based on selfish expediency? And, finally, did the Nazis explicitly embrace Haeckel's Darwinism?

Nineteenth-century Europe witnessed tremendous scientific, technological, and commercial advances, which seemed to confirm religious assumptions about signs of divine favor. The discovery of increasingly more complex fossils in ascending layers of geological formations indicated that progressive developments had been the general story of life on earth. Darwin believed his theory could explain these presumed facts of biological and social progress, since "as natural selection works solely by and for the good of each being, all corporeal and mental endowments will tend to progress towards perfection."[6] He not only thought the progressive development of individual species could be read in the fossil record but, like his disciple Haeckel, also believed that progressive advance could be detected in the developing embryo, which was left as a dynamic "picture" of the ascending morphological stages traversed in evolutionary history.[7] Darwin, too, employed the biogenetic law.

This progressivist view of animal species was consistent with the belief that the various human groups could also be arranged in a hierarchy from lower to higher. The effort to classify and evaluate the human races, however, had begun long before Darwin and Haeckel wrote. In the mid-eighteenth century, Carolus Linnaeus (1707–1778) and Johann Friedrich Blumenbach (1752–1840) first began systematically to classify human races and evaluate their attributes. In the early nineteenth century, Georges Cuvier (1769–1832), the most eminent biologist of the period, divided the human species into three varieties: the Caucasian race, the most beautiful and progressive; the Mongolian race, the civilizations of which had stagnated; and the Ethiopian race, whose members displayed a "reduced skull" and facial features of a monkey. This last group remained "barbarian."[8] That the different groups of human beings could be arranged in a hierarchy from lowest to highest was, thus, a commonplace in biology, as well as in the public mind. The U.S. Constitution recognized this kind of hierarchy when it affirmed the property rights of slave holders and stipulated that resident Africans should be counted as three-fifths of a person for purposes of deciding congressional representation.

Darwin, for his part, simply sought to explain the presumed facts of racial differences. He allowed that the human groups could be regarded either as varieties of one human species or as separate species. The decision for him was entirely arbitrary, since no real boundary could be drawn between species and varieties or races.[9] He thought it conformed better with standard usage to refer to human *races,* while Haeckel preferred to consider

different groups as distinct *species*. Though Darwin recognized
higher and lower races, he certainly did not believe this justified
less-than-humane regard for those lower in the scale. Indeed, his
abolitionist beliefs were strongly confirmed when visiting the
slave countries of South America on the *Beagle* in the early
1830s; later, he longed for the defeat of the slave-holding Southern states during the American Civil War.[10] Haeckel, on his
travels to Ceylon and Indonesia, often formed closer and more
intimate relations with natives, even members of the untouchable
classes, than with the European colonials. When incautious
scholars or blinkered fundamentalists accuse Darwin or Haeckel
of racism, they simply reveal to an astonished world that these
thinkers lived in the nineteenth century.

Gasman in a recent volume has reiterated the claim, now
widely accepted, that Haeckel's virulent anti-Semitism virtually
began the work of the Nazis: "For Haeckel, the Jews were the
original source of the decadence and morbidity of the modern
world and he sought their immediate exclusion from contemporary life and society."[11] This charge, which attempts to link
Haeckel's convictions with the Nazis' particular brand of
racism, suffers from the inconvenience of having absolutely no
foundation. The reality was quite the contrary, as is revealed
by a conversation Haeckel had in the mid-1890s on the subject
of anti-Semitism. He had been approached by the Austrian novelist and journalist Hermann Bahr, who was canvassing leading
European intellectuals on the phenomenon of anti-Semitism.
Haeckel mentioned that he had several students who were
quite anti-Semitic but that he himself had many good friends
among Jews, "admirable and excellent men," and that these ac-
quaintances had rendered him without this prejudice. He recognized nationalism as the root problem for those societies
that had not achieved the ideal of cosmopolitanism; and he did
allow that such societies might refuse entry to those who would
not conform to local customs—for instance, Russian Orthodox
Jews, not because they were Jews but because they would not

assimilate. He then offered an encomium to the educated *(gebildeten)* Jews who had always been vital to German social and intellectual life: "I hold these refined and noble Jews to be important elements in German culture. One should not forget that they have always stood bravely for enlightenment and freedom against the forces of reaction, inexhaustible opponents, as often as needed, against the obscurantists."[12] One such enlightened individual was his friend Magnus Hirschfeld (1868–1935), the physician and sexologist, who regarded Haeckel a "German spiritual hero."[13] During the Nazi period Hirschfeld had to flee for his life in the glare of his burning institute. At the turn of the century, as the black slick of anti-Semitism began to spread, Haeckel stood out for his expression of *Judenfreundschaft* (friendliness toward Jews).[14]

Perhaps the ethical proposals of a materialistic and utilitarian Darwinism have "broken with the humanitarian tradition"—in the words of one indictment—and, consequently, have sanctioned a selfish, might-makes-right kind of morality that was congenial to the Nazis.[15] Darwin, in *The Descent of Man,* did develop an explicit ethical theory based on natural selection; but he believed that his proposal overturned utilitarian selfishness and that natural selection, operating on protohuman groups, would have instilled an authentic altruism among their members.[16] Haeckel endorsed Darwin's ethical conception of altruism, which he thought a better foundation for traditional Christian morality.[17] Moreover, during the Franco-Prussian War of 1870–1871, Haeckel described a despicable phenomenon he called "military selection," in which the bravest and brightest were slaughtered on the fields of battle while the weak and cowardly were left to man the bedrooms and thereby perpetuate their low moral character. He cultivated the hope that "in the long run, the man with the most perfect understanding, not the man with the best revolver, would triumph . . . [and that] he would bequeath to his offspring the properties of brain that had promoted his victory."[18]

Despite Haeckel's being a philo-Semite and expressing an antimilitary disposition, did the Nazis yet try to recruit him—or at least his reputation, since he died a decade and a half before the Nazis came to power—and therewith embrace his Darwinism? During the 1930s, the Nazi apparatus attempted to align the new political dispensation with the views of eminent German intellectuals of earlier centuries. For instance, Alfred Rosenberg, chief party propagandist, declared Alexander von Humboldt (1769–1859), doyen of German scientists a century before, to be a supporter of the ideals of the National Socialists, even though Humboldt was a cosmopolitan friend of Jews and a homosexual.[19] Haeckel, too, was enlisted in the Nazi cause by a few ambitious academics, such as Heinz Brücher, who contended that Haeckel's evolutionary monism easily meshed with Hitler's racial attitudes.[20] But almost immediately, in the mid-1930s, the official guardians of party doctrine quashed any suggestion of consilience between Haeckel's Darwinism and the kind of biology advanced by their members. Günther Hecht, who represented the National Socialist Party's Department of Race-Politics *(Rassenpolitischen Amt der NSDAP),* issued an admonition:

> *The common position of materialistic monism is philosophically rejected completely by the völkisch-biological view of National Socialism* . . . The party and its representatives must not only reject a part of the Haeckelian conception—other parts of it have occasionally been advanced—but, more generally, every internal party dispute that involves the particulars of research and the teachings of Haeckel must cease.[21]

Kurt Hildebrandt, a political philosopher at Kiel writing in the same party organ, likewise dismissed as simply an "illusion" Haeckel's presumption that "philosophy reached its pinnacle in the mechanistic solution to the world puzzles through Darwin's descent theory."[22] These warnings were enforced by an official edict of the Saxon ministry for bookstores and libraries condemning material inappropriate for "National-Socialist formation and education in the Third Reich." Among the works to be expunged were those by "traitors," such as Albert Einstein; those by "liberal democrats," such as Heinrich Mann; literature by "all Jewish authors no matter what their sphere"; and materials by individuals advocating "the superficial scientific enlightenment of a primitive Darwinism and monism," such as Ernst Haeckel.[23]

Nazi biology formulated theories of racial degeneracy and executed a horrendous eugenic prophylaxis. But these racial notions and criminal acts were rarely connected with specific evolutionary conceptions of the transmutation of species and the animal origin of all human beings, even if the shibboleth "struggle for existence" left vaporous trails through some of the biological literature of the Third Reich. The perceived materialism of Darwinian biology and Haeckelian monism deterred those who cultivated the mystical ideal of a transcendence of will. Pseudo-scientific justifications for racism would be ubiquitous in the early twentieth century, and Hitler's own mad anti-Semitism hardly needed support from evolutionary theorists of the previous century. Weikart and Christian conservatives have attempted to trace a path from Darwin to Hitler by way of Haeckel, but their efforts must stumble against the many barriers I have noted in this chapter. While attempting to hack through an impenetrable thicket of facts, they failed to notice the great highways leading to the Third Reich that passed through the wreckages of World War I—the economic havoc, the political turmoil, and the pervasive anti-Semitic miasma created by Christian apologists. Complex historical phenomena such as the advent of the Nazi regime require complex causes to give them account—a historiographic axiom unheeded by those perpetuating the myth of Darwinian complicity in the crimes of the Nazis.

THAT THE SCOPES TRIAL ENDED IN DEFEAT
FOR ANTIEVOLUTIONISM

Edward J. Larson

The antievolutionists won the Scopes trial; yet in a more
important sense, they were defeated, overwhelmed by the tide
of cosmopolitanism.

> —William E. Leuchtenburg, *The Perils of Prosperity,*
> *1914–1932* (1958)

Of all the myths of science and religion discussed in this volume,
only one was spawned by a historical event that occurred the
United States. In 1925, Tennessee outlawed teaching the theory
of human evolution in public schools. Responding to the invita-
tion of the American Civil Liberties Union, which opposed the
statute on free-speech grounds, town leaders in Dayton, Tennessee, decided to test that new statute in court by arranging a
friendly indictment of a local science teacher named John Scopes.
Come what may, they wanted publicity for their community.
Scopes agreed to the scheme and soon hundreds of reporters descended on Dayton to cover an event that its media-savvy participants billed as "a battle-royale between science and religion."
Three-time Democratic Party presidential nominee and former
Secretary of State William Jennings Bryan, an already legendary
orator with liberal political views and conservative religious
ones, volunteered to assist the prosecution. Famed defense lawyer
and crusading secularist Clarence Darrow joined the defense

team. This much was fact: The myth took wing from near-surreal substance.[1]

With their tone set by syndicated columnist H. L. Mencken, journalists covering the trial began embellishing events in Dayton even as they unfolded. In many reports, Scopes became the victim of benighted townspeople intent on quashing religious dissent. Although a good story, these accounts made the event unduly local and personal. Tennessee lawmakers had banned the teaching of human evolution in public schools in response to a national crusade by conservative Christians. The people of Dayton had no part in that larger episode, and Scopes was not their hapless victim. Initial news reports at least got the verdict right. Scopes lost on the basis of unrebutted trial testimony that he had taught about human evolution from the state-prescribed science textbook and was fined one hundred dollars. Some articles also correctly noted that he received offers of book deals, speaking invitations, and a scholarship to the University of Chicago. Five days after his victory and six days after being subjected to a bizarre courtroom interrogation by Darrow, Bryan died in Dayton from apoplexy, a condition possibly brought on by the strenuous trial, which was conducted in oppressive heat. Media accounts at the time suggested that the trial mainly served to intensify interest and harden positions on both sides of the public controversy over teaching evolution. These contemporaneous reports were generally accurate; ultimately, however, the myth would veer more sharply from reality.

An early retelling of the Scopes trial appeared in Frederick Lewis Allen's best-selling 1931 history of the 1920s, *Only Yesterday*. Of Darrow's interrogation of Bryan, Allen wrote, "It was a savage encounter, and a tragic one for the ex-Secretary of State. He was defending what he held most dear . . . and he was being covered with humiliation." Regarding the trial itself, Allen noted, "Theoretically, Fundamentalism had won, for the law stood. Yet really Fundamentalism had lost . . . Civilized opinion everywhere had regarded the Dayton trial with amazement and amusement, and the slow drift away from Fundamentalist certainty continued."[2] These became two main elements of the Scopes myth: The trial had discredited Bryan and halted the antievolution movement. Although not true for those within the nation's growing conservative Christian subculture, it seemed so to religious liberals and secular Americans, and they repeated it to themselves.

By the 1960s, the myth had become a feature of mainstream American history textbooks. In *The American Pageant,* Stanford University historian Thomas A. Bailey commented on the Scopes trial: "The Fundamentalists at best had won only a hollow victory, for the absurdities of the trial cast ridicule on their cause. Increasing numbers of Christians found it possible to reconcile the realities of religion with the findings of modern science."[3] In their bellwether collegiate text, *History of the American Republic,* historians Samuel Eliot Morison, Henry Steele Commager, and William E. Leuchtenburg asserted: "Within a few days after his ordeal, Bryan was dead, and with him died much of the older America . . . The fundamentalist crusade, although it now had a martyr, no longer had the same force."[4] In *The American Republic,* Richard Hofstadter, William Miller, and Daniel Aaron stressed Darrow's "pitiless questioning" of Bryan. "The country's ridicule thereafter took much of the sting from fundamentalist attacks," they noted.[5] Hofstadter's Pulitzer Prize–winning book, *Anti-Intellectualism in American Life,* added in 1962, "Today the evolution controversy seems as remote as the Homeric era to intellectuals in the East."[6]

The enduring 1955 Broadway play and 1960 movie *Inherit the Wind* crystallized the modern Scopes myth. In the movie version, town officials led by a fanatical (but entirely fictional) fundamentalist minister arrest Scopes for telling his students about the Darwinian theory of human evolution. "Do we curse the man who denies the Word?" the minister asks townspeople at one point. "Yes," they reply in unison. "Do we cast out this sinner in our midst?" he adds, prompting a mightier affirmation from the crowd. "Do we call down hellfire on the man who has sinned against the Word?" he shouts. The mob roars its assent.[7] Limited to a few sets, the play begins with a jailed Scopes explaining to his girlfriend, "You know why I did it. I had the book in my hand, Hunter's *Civic Biology*. I opened it up, and read to my sophomore science class Chapter 17, Darwin's *Origin of Species*." For doing his job, Scopes "is threatened with fine and imprisonment," according to the script.[8] The ensuing trial becomes a virtual religious inquisition, with Bryan mercilessly grilling the defendant's invented girlfriend—who happens to be the minister's "pretty, but not beautiful," daughter[9]—about Scopes's religious doubts. Nothing of the sort ever happened at the actual trial. *Inherit the Wind*'s closing scenes put the final gloss on the Scopes myth: Bryan is debunked and antievolutionism discredited.

In the play and movie, Bryan's debunking culminates with his interrogation by Darrow as the defense's surprise closing witness. Bryan did not need to take the stand, of course, but he actually did so and this is where the historical account is much richer than the fictional one. "They did not come here to try this case," Bryan explained early in his *real* testimony. "They came here to try revealed religion. I am here to defend it, and they can ask me any questions they please."[10] The episode provided a ready vehicle for mythmaking.

Thinking the trial all but over, except for the much-awaited closing oratory, and hearing that cracks had appeared in the ceiling below the overcrowded, second-floor courtroom, the judge had moved the session outside, onto the courthouse lawn. When the defense then called Bryan as a witness, the crowd quickly swelled from the five hundred persons that evacuated the courtroom to an estimated three thousand people spread over the lawn—nearly twice the town's normal population. "Then began an examination which has few, if any, parallels in court history," a Tennessee newspaper reported. "In reality, it was a debate between Darrow and Bryan on Biblical history, on agnosticism and belief in revealed religion."[11] Darrow posed the well-worn questions of the village skeptic: Did Jonah live inside a whale for three days? How could Joshua lengthen the day by making the sun (rather than the earth) stand still? Did God create Eve from Adam's rib? Such questions, Darrow later explained, compelled Bryan "to choose between his crude beliefs and the common intelligence of modern times."[12] Darrow questioned Bryan as a hostile witness, peppering him with queries and giving him little chance for explanation. At times it seemed Bryan was in the firing line:

> "You claim that everything in the Bible should be literally interpreted?
>
> "I believe everything in the Bible should be accepted as it is given there; some of the Bible is given illustratively . . .
>
> "But when you read that . . . the whale swallowed Jonah . . . how do you literally interpret that?
>
> ". . . I believe in a God who can make a whale and can make a man and make both of them do what he pleases . . .
>
> "But do you believe he made them—that he made such a fish and it was big enough to swallow Jonah?
>
> "Yes sir. Let me add: One miracle is just as easy to believe as another.
>
> "It is for me . . . just as hard.
>
> "It is hard to believe for you, but easy for me . . . When you get beyond what man can do, you get within the realm of miracles; and it is just as easy to believe the miracle of Jonah as any other miracle in the Bible."[13]

Such affirmations surely sounded quaint to many twentieth-century Americans, but they accorded with the faith of millions. Bryan did concede various points of biblical interpretation generally accepted by conservative Christians of his day. Deferring to Copernican astronomy, for example, Bryan suggested that God extended the day for Joshua by stopping the earth rather than the sun. Similarly, in line with nineteenth-century evangelical thought, Bryan affirmed that the Genesis days of creation represented long periods of time and that the universe was untold millions of years old, leading to the following exchange, with Darrow asking the questions:

> "Have you any idea of the length of these periods?
> "No; I don't.
> "Do you think the sun was made on the fourth day?
> "Yes.
> "And they had evening and morning without the sun?
> "I am simply saying it is a period.
> "They had evening and morning for four periods without the sun, do you think?
> "I believe in creation as there told, and if I am not able to explain it I will accept it."[14]

Darrow never directly asked Bryan about the theory of evolution. He knew that, given the opportunity, Bryan would respond with glib answers about alleged gaps in the fossil record and the supposedly adverse social consequences of Darwinian thinking. Tell students that they descended from lower animals and they will act like apes, Bryan would argue. Portray them as made in God's image, and they might behave more like angels.

As Darrow pushed his various lines of questioning, Bryan increasingly admitted that he simply did not know the answers. He had no fixed idea of what would happen to the earth if it stopped for Joshua, or about the antiquity of human civilization, or even regarding the age of the earth. "Did you ever discover where Cain got his wife?" Darrow asked. "No sir; I leave the agnostics to hunt for her," Bryan defiantly replied.[15] The crowd warmly cheered on Bryan. Darrow received little applause but inflicted the most jabs. "The only purpose Mr. Darrow has is to slur the Bible, but I will answer his questions," Bryan exclaimed near the end. "I object to your statement," Darrow shouted back. "I am examining your fool ideas that no intelligent Christian on earth believes."[16] The judge had heard enough. Over two hours after Bryan had taken the stand, the judge abruptly adjourned court and never allowed the interrogation to resume.

Inherit the Wind reconstructed this storied encounter. In the play and the movie, Bryan assails evolution solely on narrow biblical grounds and denounces all science as "Godless," rather than only the so-called false science of evolution.[17] Instead of acknowledging the "day-age" interpretation of the Genesis account, the Bryan of *Inherit the Wind* maintains on alleged biblical authority that God created the universe in six days beginning "on the 23rd of October in the Year 4,004 B.C. at—uh, at 9 A.M.!"[18] By the end of his stage testimony, Bryan is totally befuddled. "Mother," he pleads to his wife. "They are laughing at me, Mother!"[19]

The writers portray Bryan's hold on the crowd gradually slipping away during his testimony and then being broken altogether a day later when he objects in court to the small size of the penalty imposed on Scopes. Even though the real Bryan originally recommended that Tennessee's antievolution statute carry no criminal penalty and once offered to pay Scopes's fine himself, the fictional Bryan instead protests, "Where the issues are so titanic, the court must mete out more drastic punishment . . . to make an example of this transgressor!"[20] The other actors ignore Bryan when he attempts, after the trial, to deliver a courtroom rant against the theory of evolution, precipitating his fatal collapse. He is carried off incoherently delivering a presidential inaugural address. "The mighty Evolution Law explodes with a pale puff of a wet firecracker," the stage directions explain.[21] To ensure that viewers appreciate this point, the writers have Scopes ask Darrow after the jury convicts him, "Did I win or did I lose?" Darrow answers, "You won . . . Millions of people will say you won. They'll read in their papers tonight that you smashed a bad law. You made it a joke!"[22] When Darrow hears that Bryan has died, this version of the myth famously has Darrow quote from the Bible, "He that troubleth his own house shall inherit the wind: and the fool shall be servant to the wise in heart."[23] Meanwhile, the trial has emboldened the minister's daughter to forsake her father and leave town with Scopes. Alone in the courtroom at the end, Darrow picks up the defendant's copy of the *Origin of Species* and the judge's Bible. After "balancing them thoughtfully, as if his hands were scales," the staging directions state, the attorney "jams them in his briefcase, side by side," and walks off quietly.[24] The message was clear: Even in Dayton, the Scopes trial ended in defeat for Bryan and his brand of Bible-based antievolutionism, but a triumph for the best in science *and* religion. Indeed, *Inherit the Wind* remakes the trial into a case for religious tolerance and portrays ennobled science and enlightened Christianity as comfortably compatible.

When the trial occurred, people generally did not see it this way. In its immediate aftermath, for example, most newspaper editorialists depicted the trial as inconclusive and predicted that the antievolution controversy would intensify as a result. When Bryan died unexpectedly in Dayton shortly after the trial, he was lionized by millions. Crowds lined the railroad track as a special train carried his body to Washington for burial at Arlington National Cemetery. Thousands filed by the open casket, first in Dayton, then in several major cities along the train route, and finally in the nation's capital. America's political elite attended the funeral, with senators and cabinet members serving as pallbearers. Country music ballads picked up the lament while fundamentalist leaders competed to carry on Bryan's crusade against teaching evolution. Several states and countless local school districts responded to the trial and Bryan's death by imposing their own restrictions on teaching evolution, especially after 1927, when the Tennessee Supreme Court upheld its state law as constitutional. Months after the trial ended, Mencken sneered about Bryan, "His place in Tennessee hagiocracy is secure. If the village barber saved any of his hair, then it is curing gallstones down there today."[25] An antievolutionist college named for Bryan soon opened in Dayton and has expanded over time.

The growth of Bryan College tracked larger developments within the American church. Protestantism's conservative, antievolution wing has expanded at the expense of its mainstream center and modernist wing. Creation science, committed to advancing scientific arguments for a more literal reading of Genesis than even Bryan countenanced, has taken root among Fundamentalists and Pentecostals. By the dawn of the twenty-first century, public-opinion surveys indicated that fully half of all Americans affirm that God separately created the first humans within the past ten thousand years, with an even greater proportion of them supporting the inclusion of creationist ideas in public-school biology courses. The years since the Scopes trial are pockmarked with efforts at the state and local levels to limit the teaching of evolution in American public schools. Antievolutionism did not die in Dayton; Bryan's crusade continues.

MYTH 21

THAT EINSTEIN BELIEVED IN A PERSONAL GOD

Matthew Stanley

Einstein saw his entire vocation—understanding the workings of
the universe—as an attempt to understand the mind of God.

—Charles Krauthammer, *Washington Post* (2005)

Einstein's belief in an intelligent designer thus derived not from a
pre-conceived religious bias, but from the phenomenal insights
into the Universe that he possessed as the most brilliant scientist
who ever lived. His recognition of a creator refutes the recent
claims by atheists that belief in any sort of god is unscientific.

—Stephen Caesar, "Investigating Origins: Einstein and
Intelligent Design" (2007)

A curious urban legend has been appearing regularly in email
inboxes for several years now. In it, an atheist professor tries
mightily to prove that there is no God, only to be calmly refuted
at each turn by a courageous student. The devout agent of his
defeat is identified as none other than the most recognizable scientist of the modern age—Albert Einstein (1879–1955).[1]
    The legend—completely false—seeks to marshal the authority
of Einstein for the existence of God. At first glance he seems a
good fit for this story. On several occasions he identified himself
as being a religious person, and he was frequently given to sage-
like pronouncements about the actions and intentions of God. But
he was also quite clear that he completely rejected the kind of personal God (a God with recognizably humanlike characteristics,

who answers prayers and watches over creation) defended by the
fictional student in the urban legend. Further, he had little interest
in religious traditions or orthodox systems of belief.

So what is going on here? How could Einstein dismiss a personal God and virtually the entire Western theistic tradition while
still claiming to be religious? The answer lies in what he saw as
the nature of religion, his personal religiosity, and his views on
the appropriate relationship between religion and science.

Einstein's rejection of a personal God was based on ideas that
have troubled theologians for centuries. He could not bring himself to accept a benevolent, omniscient, omnipotent God who
ruled over a world filled with evil and suffering. He also felt that
the existence of such a deity would remove the need for personal
responsibility. But the crux of his argument was that the physical laws so diligently and impressively assembled by generations
of scientists seemed to leave no space for divine action:

> The more a man is imbued with the ordered regularity of all events
> the firmer becomes his conviction that there is no room left by the
> side of this ordered regularity for causes of a different nature. For
> him neither the rule of human nor the rule of divine will exists as an
> independent cause of natural events.[2]

The laws of nature, to Einstein, were completely causal. That is,
every event in the physical world was caused by another physical
event, and this cause was described accurately by scientific laws.
This "regularity" explained, at least in principle, every occur-
rence and phenomenon in the universe. There was no empty
space of causation in which the hand of God could have acted.

He admitted that an intervening God could never be dis-
proven as long as there remained areas that science did not
understand precisely, but he vigorously defended a world view
built only on unshakable physical laws:

> I cannot prove to you that there is no personal God, but if I were to
> speak of him, I would be a liar. I do not believe in the God of theology
> who rewards good and punishes evil. My God created laws that take

care of that. His universe is not ruled by wishful thinking, but by immutable laws.[3]

To Einstein, divine judgment and the efficacy of prayer seemed completely implausible in light of the consistency of science.

Einstein did not see this as a simple conflict between science and religion, in which laws of science destroyed the notion of a personal God. Rather, he thought that moving beyond this concept was simply a stage in the maturation of religion. He postulated that the origins of religion could be found in the existential fears of primitive man, which drove him to create "illusory beings" that later developed into an intervening, anthropomorphic God.[4]

The ultimate stage of religion was one in which this anthropomorphic God was discarded in favor of "cosmic religious feeling." Einstein said all the great religious leaders in history (among whom he included Democritus, Francis of Assisi, and Baruch Spinoza) were operating in this stage, in which the "individual feels the futility of human desires and aims and the sublimity and marvelous order which reveal themselves both in nature and in the world of thought . . . [H]e wants to experience the universe as a single significant whole." Religions evolved to this point had no dogma, central church, or personal God. It is at this point that science has a role to play, by "purifying the religious impulse of the dross of its anthropomorphism" through the construction of the causal network that makes a personal God impossible. This view of God is clearly incompatible with the basic assumptions of Western monotheism. Indeed, Einstein's vision of the final stage of religious evolution even classifies a number of traditional beliefs and practices as being nonreligion: "As long as you pray to God and ask him for some benefit, you are not a religious man."[5]

Along with an anthropomorphic God Einstein also wanted to discard the idea that such an entity was necessary for moral behavior. Morality, he argued, was an "exclusively human concern with no superhuman authority behind it." A postulated divine lawgiver would give no extra weight to morality; it could even be counterproductive: "The foundation of morality should not be made dependent on myth nor tied to any authority lest doubt about the myth or about the legitimacy of the authority imperil the foundation of sound judgment and action."[6] In the end, Einstein rejected a personal God as implausible, primitive, and even dangerous to humanity.

If Einstein was so hostile to a personal deity, what do we make of his frequent invocation of God in conversation? He once described his goal in science this way: "I want to know how God created this world. I'm not interested in this or that phenomenon, in the spectrum of this or that element. I want to know His thoughts, the rest are details."[7] In debates over quantum mechanics in the 1920s and 1930s, he made pronouncements such as "God does not play dice" frequently enough to exasperate his colleagues. How does Einstein's ease with this kind of language fit with his rejection of a personal God?

Taken in context, these comments were almost certainly metaphorical, as became apparent when Einstein clarified his ideas. He was once asked exactly what he meant when he said, "Subtle is the Lord, but malicious He is not." He replied, "Nature hides her secret because of her essential loftiness, but not by means of ruse." The "Lord" of the first version is easily transformed into "Nature" in the second, suggesting a far less supernatural interpretation of his meaning. Similarly, consider a comment made to his assistant: "What I am really interested in is knowing whether God could have created the world in a different way; in other words, whether the requirement of logical simplicity admits a margin of freedom." As the physicist-historian Max Jammer points out, the "in other words" clearly shows that the reference to God in the first part was simply rhetorical.[8] In almost all of these cases, the term *God* appears to have functioned as a kind of linguistic placeholder for Einstein, providing an evocative and memorable way to refer to the orderliness and comprehensibility

of the universe. Of course, it was exactly this orderliness that un-
derlay his cosmic religious feeling, but it bears little resemblance
to traditional notions of God.

Einstein's rejection of conventional theology is clear, but how
then are we to relate that rejection to his claimed religiosity?
Fortunately, he does this very clearly for us in a letter to a correspondent surprised to hear the great professor described as a religious believer:

> It was, of course, a lie what you read about my religious convictions,
> a lie which is being systematically repeated. I do not believe in a per-
> sonal God and I have never denied this but have expressed it clearly.
> If there is something in me which can be called religious then it is the
> unbounded admiration for the structure of the world so far as our
> science can reveal it.

The confusion comes from exactly what Einstein meant by *religious.* Einstein's religiosity shared very little with conventional
religion. By his own account he was raised by "entirely irreligious (Jewish) parents," but as a child he briefly indulged in traditional religiosity of his own accord. This ended at age twelve,
when reading popular science books convinced him that the stories of the Bible were unlikely to be true. He did credit that brief
period of religious fervor with inspiring in him an urge to push
beyond the "merely personal" to something greater.[9]

This attitude was the root of what Einstein meant by religion.
He described a religious person as one who has "liberated himself
from the fetters of his selfish desires and is preoccupied with
thoughts, feelings, and aspirations to which he clings because of
their superpersonal value." Note that this description makes no
reference to a God, supernaturalism, scriptures, or religious practices or communities. Many of his correspondents objected to
calling this *religion,* but Einstein insisted that it was the best term
available:

> I can well understand your aversion to the use of the word "religion"
> when what is meant is an emotional or psychological attitude, which

is most obvious in Spinoza. I have found no better expression than "religious" for confidence in the rational nature of reality insofar as it is accessible to human reason . . . For all I care, the parsons can make capital of it.

His notion of religion was founded not on divine revelation or action but on his awe that the world was comprehensible. When famously asked to answer the query "Do you believe in God?" by telegram in fifty words or less, he replied: "I believe in Spinoza's God, Who reveals Himself in the lawful harmony of the world, not in a God Who concerns Himself with the fate and doings of mankind."[10] Einstein repeatedly allied himself with Spinoza's (1632–1677) religious outlook, which he interpreted as presenting a pantheistic God completely lacking in personality or will and manifest only in the orderly nature of the cosmos. It is critical to understand Einstein's occasional references to a superior "intelligence" in the universe in this context.[11] Spinoza's God certainly possessed intelligence in some sense, but it is explicitly not the intelligence of a being that has will and can cause actions. In this kind of monistic theology, there is no separation between God and the world, so expecting the deity to direct miracles or interventions becomes nonsensical. God *is* nature and its laws, rather than being the creator of them. Spinoza appears to be one of Einstein's major influences in thinking about religion and God, and we can see that Dutch philosopher's work echoed strongly in Einstein's insistence on strict determinism in the world. Both argued that the foundation of religion was in the rationality, order, and comprehensibility of the world, not in the miraculous acts of an anthropomorphic deity.

As unusual as Einstein's conception of religion was, he took it seriously. Lacking an inspirational God, he grounded his religiosity on the aforementioned cosmic religious feeling. This was a state in which the "individual feels the futility of human desires and aims and the sublimity and marvelous order which reveal themselves both in nature and in the world of thought."

Einstein said this was what truly religious people moved on to after discarding a personal God. This "rapturous amazement" at the mysterious order of the universe was also the "guiding principle" of the work of scientists.[12] Faith, then, was critical to the advancement of science. But it was faith in rationality, not divine intervention.

Interestingly, Einstein's religiosity had the same foundation as his rejection of a personal God: the idea of universal causality. The same cosmic orderliness that provided rapturous amazement allowed no room for a deity. This cosmic religious feeling most powerfully manifested itself not in traditional religion but in the practice of science. "In this way the pursuit of science leads to a religious feeling of a special sort, which is surely quite different from the religiosity of someone more naïve." This, Einstein said, resolved any idea of a conflict between science and religion. Science depended on religion for faith in a comprehensible universe and religion depended on science for the discovery of the awesome order of the universe. These notions are behind his aphorism that "science without religion is lame, religion without science is blind."[13] And while science and religion shared a kindred outlook on the universe, Einstein was adamant that there was no overlap between their content. When asked what implications relativity had for religion he once replied, "None. Relativity is a purely scientific matter and has nothing to do with religion."[14] The influence of religion on science and vice versa was solely a matter of a reverent attitude toward the cosmos. Einstein felt that in an important way this meant modern scientists had displaced the religious leaders of old. Since science was the best source of cosmic religious feeling, "serious scientific workers are the only profoundly religious people" of the twentieth century.[15]

We can now understand why Einstein called himself religious. Indeed, he did not care for atheism and bristled at the suggestion that scientists naturally fell into that camp.[16] When sent an atheistic book, he protested that the arguments in it applied only to a personal God. But it is important to remember that Einstein's religiosity was quite different from orthodox religion. In addition to denying a personal God, he rejected such foundations of Western theism as life after death and human free will.[17] He was also fairly hostile to organized religion and its institutions. He once wrote that the fantasies of primitive man brought "boundless suffering" once they had "crystallized into organized religion." He saw the influence of religious institutions as primarily negative and as only being perpetuated by indoctrinating children "by way of the traditional education machine."[18]

Einstein's rejection of organized religion might seem strange in light of his public support for a Jewish homeland. Again, the apparent inconsistency arises from his idiosyncratic understanding of what it meant to be Jewish. His embrace of a Jewish identity came fairly late in life; he declared he had no religious affiliation until forced to choose one by Austro-Hungarian bureaucracy. He was never bar mitzvahed and never attended Jewish services; he owned a pair of phylacteries but never used them. Einstein did celebrate the ethical legacy of the Hebrew tradition, but for the most part he thought being a Jew meant being part of a community held together by "blood and tradition, and not of religion only." He supported Zionism as a way for an oppressed group to find survival and safety. He was spurred toward such thinking by the anti-Semitic attacks directed against him by the Nazis, who accused Einstein of having corrupted science with his "Talmudic thinking." These attacks, leveled against someone completely uninterested in scriptures or the dogmas associated with them, would have been laughable if they had not been so dangerous. Einstein repeatedly said that Zionism should be supported as a way for Jews to defend themselves and not for any scriptural or theological reason. Indeed, he went so far as to question whether Judaism was a religion at all:

Judaism is not a creed . . . it is concerned with life as we live it and we can, to a certain extent, grasp it, and nothing else. It seems to me, therefore, doubtful whether it can be called a religion in the accepted sense of the word, particularly as no "faith" but the sanctification of life in a suprapersonal sense is demanded of the Jew.[19]

As with his identity as a "deeply religious" person, his identity as a Jew depended on his own peculiar set of definitions.

THAT QUANTUM PHYSICS DEMONSTRATED

THE DOCTRINE OF FREE WILL

Daniel Patrick Thurs

Moralists have not been slow to draw the conclusion from the work of [Werner] Heisenberg and others that since determinism (that is, materialism and mechanism) goes by the board we may talk of "free will" again. The argument has been punctured over and over again by Heisenberg, Neils Bohr and the late Sir Arthur Eddington.

—Waldemar Kaempffert, *New York Times*

[T]he field currently seems divided between those who fear that quantum mechanics may validate the existence of the mind or free will and those who hope that it validates channeling or aliens.

—Denyse O'Leary, "The ID Report"

Unlike many of the myths elsewhere in this book, this one— namely, the notion that quantum mechanics provided scientific justification for a belief in free will—is diffuse. It has not had time to take on the patina of established mythology, in part because claims about the relationship between quantum theory and religion are comparatively recent. Nor has it, like some other modern myths, had the advantage of a famous celebrity (such as Einstein) or a dramatic event (such as the Scopes Trial) to carry it into public view. And it deals with a somewhat obscure realm of science, one not taught in primary or secondary schools. There is therefore less at stake in terms of social or cognitive authority in its retelling or in its denial. Finally, this myth does not deliver much in the way of exciting conflict. Rather than displaying the warfare between science and religion in gory detail, most claims about the religious implications of quantum ideas seek, in the words of physicist and author Amit Goswami, to "bridge the age-old gap between science and spirituality."[1]

Nevertheless, the notion that quantum mechanics has been used to demonstrate the existence of free will circulates in print, on film, and especially on the Internet. And its spread is due to a very good reason. It is, in one sense, absolutely true. A wide range of spiritually minded thinkers from scientists and philosophers to priests and mystics have put quantum theory to work justifying free will, as well as a mixed bag of other religious views. The legitimacy of such use is, however, a question of a different order. It is a topic of debate, albeit one between groups and individuals with divergent relationships to orthodoxy. New Age "quantum mystics," some with training in the sciences but tenuous relationships with the scientific community, routinely spin out the metaphysical implications of quantum theory while skeptics, frequently with many ties to scientific institutions, denounce "quantum quackery." If the historian *as historian* has any role, it is to expose the roots of such controversy rather than to leap into the fray and parrot the arguments of one side or another.

Such a view undoubtedly does little to satisfy the desire to see one more myth meet its end. By way of consolation, I can offer up two related claims for sacrifice, both of which reflect widespread truisms about science and religion. Quantum theologians often assert that their conclusions are intrinsic to the theory, frequently drawing on a sense that science and religion share some essence laid most bare by quantum mechanics. Alternatively, skeptics typically stress the cleavage between science and religion and therefore present all attempts to mix the two as fuzzyheaded impositions on sober science. The historical record suggests a more complex situation. The religious revelations of the new physics have had much more to do with the contexts in which they appeared than with the theory itself. In that sense, such interpretations *were* impositions, but the same might be said of any interpretation of a mathematical equation. Moreover, such impositions were not purely the province of scientific outsiders. The founders of quantum theory were guilty of their share of philosophical and religious speculations. To see the full import of those speculations, it is important to say a few words first about quantum mechanics itself and second about the more general cultural discussion that was already well underway when quantum's founders made their own views known.

As in the case of evolution, religious discussion of quantum theory has tended to focus on the implications of randomness. The major vehicle of that randomness is the theorem popularly known as Heisenberg's uncertainty principle, which asserts the impossibility of determining, among other things, the position and momentum of atomic particles simultaneously. Beginning in the 1920s, Niels Bohr (1885–1962), Werner Heisenberg (1901–1976), and others placed the uncertainty principle at the very heart of the so-called Copenhagen interpretation of quantum physics, depicting the microcosm as a composition of jostling and overlapping possibilities, all contained in a system's wave function. Observation prompted that wave function to collapse through an instantaneous, though mysterious, quantum "jump" into a single actuality. Other interpretations have also appeared, particularly during the 1950s, including Hugh Everett's (1930–1982) many-worlds interpretation and David Bohm's (1917–1992) deterministic reworking of quantum ideas. But because both tended to explain away the role of chance in the universe rather than take the statistical bull by the horns, neither attracted much religious attention.[2]

The first attempts to make religious hay of the Copenhagen interpretation did not wait for long. The leading light of these early efforts was British astronomer Arthur Stanley Eddington

(1882–1944).[3] Eddington's *Nature of the Physical World,* which appeared in 1928, was based on a series of lectures that took place a year earlier and a few months before Heisenberg's publication of his uncertainty principle.[4] In this work, Eddington, a devout Quaker with tendencies toward mysticism, offered one of the earliest surveys of quantum theory generally and ended with an extended discussion of its philosophical and religious implications. He continued to address the relationship between modern physics and religion in other works into the 1930s, when he was joined by fellow astronomer Sir James Jeans (1877–1946), the American physicist Arthur H. Compton (1892–1962), and a variety of lesser-known popular writers.

The major issues with which these authors grappled included the viability of materialism, the possibility of human free will, and the ability of God to intervene in the natural world. Compton, who won the Nobel Prize in physics in 1927, was particularly outspoken about the last two, claiming that the "amazing world of the atom points . . . to the idea that there is a God."[5] Likewise, he used the uncertainty principle to argue in favor of free will. Eddington, also motioning toward the uncertainty principle, took a swipe at free will in *Nature of the Physical World,* though his treatment of the topic was somewhat offhand. Indeed, the real thrust of this first wave of quantum speculation was not the creation of a new natural theology—this was certainly not Eddington's primary concern—but rather the more modest claim that Victorian dreams of a purely materialistic universe had finally dissolved. Rather than directly supporting religion, the acceptance of uncertainty as a scientific principle helped to make room for spirituality in the modern world.

Not everyone was happy with such conclusions. In her *Philosophy and the Physicists* (1931), the philosopher L. Susan Stebbing (1885–1943) took Eddington to task for a variety of philosophical mistakes. Likewise, the philosopher William Savery (1875–1945) and the physicist Charles G. Darwin (1887–1962) criticized Eddington's attempt to apply the uncertainty principle to the issue of free will.[6] But the retreat from what many people perceived as the "old and happily obsolete Victorian cocksuredness" did appeal to some, particularly in the United States, where the antievolution movement had crested in the late 1920s, and where the humbling impact of the Great Depression, which some people blamed on the too-relentless advance of science and technology, was being felt.[7] An American magazine correspondent claimed in 1929 that if quantum theory was true, then it restored the "hope for something beyond."[8]

Speculations over the meaning of quantum physics were hardly sweeping through popular culture, however. Quantum theory was often eclipsed by relativity. In contrast to the literary machine that produced a number of popular accounts of relativity, there were few popularizations of quantum ideas until after World War II—with the notable exception of George Gamow's *Mr. Tompkins in Wonderland* (1940). Likewise, during the 1920s and 1930s, most public discussion of science and religion not monopolized by the debate over evolution revolved around Einstein's remarks about God and deep mysteries. By the 1950s, quantum physics and its philosophical implications were receiving more attention. The first hints of a massive new wave of interest in quantum theory's religious implications washed onto shore in 1975, when physicist Fritjof Capra (b. 1939) published *The Tao of Physics*. This book detailed what Capra claimed were myriad connections between modern physics, particularly quantum mechanics, and Eastern mysticism.

Earlier writers, such as C. H. Hsieh, author of *Quantum Physics and the I Ching* (1937), had made similar connections.[9] But Capra's work brought such ideas to a mass audience. Gary Zukav's *The Dancing Wu Li Masters* (1979), a similar overview of quantum physics in the context of Eastern philosophy, appeared on its heels. By the 1990s, the amalgamation of quantum mechanics and Eastern mysticism had been taken up by the New Age movement. Invocations of quantum physics peppered the rhetoric of healers and prophets, such as Deepak Chopra and

Ramtha, the spiritual entity supposedly channeled by the American medium Judy Zebra Knight and the inspiration for the 2004 film *What the (Bleep) Do We Know?* Eventually notions of links between quantum mechanics and Eastern thought flowed back to the East. Tenzin Gyatso, the fourteenth Dalai Lama (b. 1935), has been particularly interested in the new physics, even introducing workshops on quantum theory into Tibetan monastic colleges.[10]

The concerns that dominated this second wave of religious interest in quantum physics were very different from those that circulated in the 1930s. In Buddhist, Taoist, and Hindu writings, God and free will received little attention. Additionally, Capra and those who followed him were not content to show how quantum theory made room for religion; rather, they sought to demonstrate the positive convergence of modern science and mystical ideas. Advocates of an Easternized quantum mechanics argued for the universal interconnection of events by appealing to the entanglement of wave functions describing different parts of larger systems, even if no single line of physical causation linked them. Likewise, they routinely stressed the theory's supposed weakening, if not complete erasure, of the line between subjects and observers. This notion reached its most extreme form in what Amit Goswami called the "Self-Aware Universe"— that is, the idea that human consciousness was responsible for the collapse of the wave function and the creation of reality.[11]

The timing of these connections was ideal. Capra's work harmonized with Western countercultural movements and their attraction to all things esoteric. The 1970s also witnessed shifts in psychology toward attempts to justify the serious study of consciousness, as well as increased curiosity about such phenomena as telepathy.[12] By the 1980s and 1990s, ideas about the powers of the human mind fit nicely with the New Age and self-help movements. A recent website devoted to "creating consciously" in life proclaimed that quantum-mechanical studies had "discovered that the thoughts and expectations of the experimenter were actually causing the experiment's outcome!"[13] Quantum mystics have also come to rely on the increased cultural capital of strangeness since the late 1960s and 1970s to question traditional beliefs. Quantum weirdness has become a powerful means of promoting enlightenment, demonstrating the illusory quality of perceived reality, and breaching established divisions between science and religion.

In so doing, quantum theologians have drawn criticism from the host of skeptics who emerged after World War II to combat pseudoscience in its myriad forms, from creationism to astrology. The physicist Victor Stenger (b. 1935) has been a particularly vocal critic of quantum mysticism.[14] In recent years the mystical interpreters of quantum physics have also had to share the stage with various Christian apologists, eager to fill the space cleared out by quantum ideas with positive support for religious belief. In 1988, the physicist-turned-theologian Robert J. Russell (b. 1946) proposed that the randomness of the microcosm provided a gap through which God could act on the natural world, an idea taken up by other Christian theologians and philosophers.[15] A Westernized interpretation of the religious implications of quantum physics has also retained a popular profile through such works as Paul Davies's *God and the New Physics* (1984), though it has not approached the cultural influence of its mystical alternatives.

These examples should make it clear that religious interpretations of the new physics have varied considerably with the context in which they appeared. But has such interpretation been a violation of its scientific core? To decide, we might look to the founders of quantum theory—those who presumably knew it best—to find a more purified vision of the theory's scope and implications. Figures such as Bohr, Heisenberg, and Wolfgang Pauli (1900–1958) have certainly figured in debate about the true implications of the theory. Advocates of metaphysical interpretations have juxtaposed the writings of such quantum pioneers with passages from the Bible, the wisdom of the Buddha, and the observations of the *I Ching* to show their uncanny similarity.[16] Needless to say, critics have offered their own correctives. In a recent biography of Bohr, Abraham Pais repeatedly stressed what he claimed was Bohr's lack of interest in philosophy and Eastern mysticism.[17]

To be sure, the founders of quantum mechanics were much more concerned with the scientific and technical details of their work than its metaphysical aspects, particularly during the heady days of the 1920s when the quantum theory was developing rapidly. Nevertheless, they did have something to say about the philosophical and even religious applications of their ideas. If Heisenberg's recollections from 1971 can be trusted, he, Bohr, Pauli, and Paul Dirac (1902–1984) were exchanging views among themselves on the relationship between science and religion as early as 1927.[18] It was not until after World War II that they generated many philosophical writings, and then sometimes in a haphazard and incomplete way. While Bohr tried to apply his principle of complementarity—the notion that reality could be described only by the juxtaposition of multiple, even contradictory, ways of viewing the world—to the problem of free will, neither he nor any of the other founders clearly invoked the uncertainty principle on the matter.[19]

Many did echo the discomfort with pure materialism that was a major theme in early religious interpretation of quantum theory. Heisenberg, who had been inspired to take up science by an early encounter with the idealistic philosophy of Plato, felt that "the Copenhagen interpretation of quantum theory has led the physicists far away from the simple materialistic views that prevailed in the natural science of the nineteenth century."[20] He also stressed a humble image science, remarking that "what we observe is not nature itself but nature exposed to our method of questioning" and, noting in an essay published in English in 1971 that physics was "undergoing a basic change, the most characteristic trait of which is a return to its original self-limitation."[21] Bohr may also have been ready to extend his ideas about complementarity to the relationship between science and religion in a way that recognized them both as necessary perspectives on the world.[22] Pauli, as Heisenberg once recalled, predicted that an adversarial relationship between knowledge and faith was "bound to end in tears."[23]

In other areas, their claims were not always as consistent and ironclad as later generations of skeptics might have wished. Both Bohr and Heisenberg explicitly denied that the observer had a role in creating reality. In *Physics and Philosophy* (1958), Heisenberg cautioned that "the introduction of the observer must not be misunderstood to imply that some kind of subjective features are to be brought into the description of nature" and noted that "the observer has, rather, only the function of registering decisions, i.e., processes in space and time, and it does not matter whether the observer is an apparatus or a human being."[24] At the same time, he frequently made statements that seemed to suggest that the old notion of an objective universe had been put to rest by quantum physics. He emphasized "a subjective element in the description of atomic events" and seconded the "old wisdom" that "when searching for harmony in life one must never forget that in the drama of existence we are ourselves both players and spectators."[25]

Finally, many of the leading figures of quantum mechanics strayed at least a little into religious heterodoxy. And sometimes they brought their science with them. The notion that human consciousness made wave functions collapse was first suggested not by a New Age guru but by Nobel Prize–winning physicist Eugene Wigner (1902–1995) in 1967, though he soon retracted his suggestion.[26] A concern with the human mind also occupied Wolfgang Pauli, who inaugurated a thirty-six-year correspondence with Carl Jung in 1932, spanning topics from Eastern mysticism to UFOs.[27] Bohr and Heisenberg each had brushes with broadly Eastern ideas as well.[28] Upon his Danish knighthood in 1947, Bohr chose a crest that included the yin-yang symbol as a sign of complementarity. Heisenberg noted explicitly that Japanese contributions to physics "may be an indication of a certain relationship between philosophical ideas in the tradition of the Far East and the philosophical substance of the quantum theory."[29]

Heisenberg's student Carl von Weizsäcker (1912–2007), who instructed the current Dalai Lama in quantum theory, apparently believed that his mentor would have been "excited to hear of the clear, resonant parallels between Buddhist philosophy and" modern physics.[30] Weizsäcker almost certainly overstated Heisenberg's investment in Eastern philosophy, as well as his interest in philosophy generally. Rather, he was himself trading on the lingering notion that there is some fundamental connection between quantum physics (and its founders) and religious or metaphysical ideas. Without serious qualification, that is clearly not true. But then, neither is the assertion of a pristine quantum theory on which skeptics and myth-busters can rely. Despite the widespread notion that science and religion are entirely distinct, at least when they are not in open warfare, their historical relationship, even in the modern world, has been much more complex. A clear view of the ongoing interpretation of quantum ideas provides an important corrective to both overenthusiastic depictions of warfare and harmony, and a valuable field in which to gain greater insight into the subtle interactions between scientific and religious concepts during the twentieth century.

THAT "INTELLIGENT DESIGN" REPRESENTS
A SCIENTIFIC CHALLENGE TO EVOLUTION

Michael Ruse

The result is so unambiguous and so significant that it must be
ranked as one of the greatest achievements in the history of science.
The discovery [of intelligent design] rivals those of Newton and
Einstein, Lavoisier and Schroedinger, Pasteur and Darwin.

  —Michael J. Behe, *Darwin's Black Box* (1996)

"Both sides [intelligent design and evolution] ought to be
properly taught . . . so people can understand what the debate is
about. Part of education is to expose people to different schools
of thought . . . You're asking me whether or not people ought
to be exposed to different ideas, and the answer is yes."

  —George W. Bush, talking to a group of journalists (2005)

Intelligent design is a winner in the public debate over biological
origins not only because it has the backing of powerful ideas,
arguments, and evidence but also because it does not turn this
debate into a Bible-science controversy. Intelligent design, unlike
creationism, is a science in its own right and can stand on its
own feet.

  —William A. Dembski, "Why President Bush Got It Right
  about Intelligent Design" (2005)

We need to answer two questions: What is intelligent design
(ID), and is it science? Answering the first, the claim is that in the
history of life on this planet, at some point or points, an intelli-

gence intervened to move things along. This was necessary, argue ID theorists, because life shows "irreducible complexity,"
and blind law—especially the Darwinian evolutionary theory
that depends on natural selection—cannot explain such complexity. Only an intelligence is able to do this.

The phrase *intelligent design* began circulating after the U.S.
Supreme Court ruled in 1987 that it was unconstitutional to require the teaching of "creation science" but allowed for the possibility of voluntarily teaching alternatives to evolution—if done
for secular reasons. Hoping to cash in on this opportunity, a
small Christian organization in Texas brought out an antievolution textbook, *Of Pandas and People: The Central Question of
Biological Origins* (1989), written by the creationists Dean H.
Kenyon, a biologist at San Francisco State University, and Percival Davis, a community-college biology teacher in Florida.
Kenyon and Davis had originally intended to write a creationist
text suitable for public schools, called *Biology and Creation*. But
when the Supreme Court decided against creation science, they
quickly replaced the words "creation" and "creationists" with
"intelligent design" and "design proponents" and gave their
work its new title. In print Kenyon and Davis defined ID as a
frame of reference that "locates the origin of new organisms in
an immaterial cause: in a blue-print, a plan, a pattern, devised by
an intelligent agent"; privately one of their collaborators called
it a "politically correct way to refer to God."[1]

ID attracted little attention before the early 1990s, when a
Berkeley law professor, Phillip E. Johnson (b. 1940), published a
short book, *Darwin on Trial* (1991). Highly critical of modern
evolutionary thought, Johnson trotted out arguments that have
been familiar since the time of the *Origin of Species*—the inade-
quacy of natural selection, gaps in the fossil record, the problem
of life's origins, and so forth. Familiar though the arguments
may have been, Johnson knew how to mount a prosecution, and
it soon became clear that his book was a more polished work
than the run-of-the-mill antievolutionist tract. By the time he

had finished, Johnson felt justified in finding the old evolutionist guilty and led him away in chains.[2]

As much as Johnson disliked Darwin and evolution in particular, he saw their widespread acceptance as symptomatic of a broader embrace of naturalism—or, as it was coming to be called, methodological naturalism—which unfairly biased science against even considering theistic explanations. "Naturalism rules the secular academic world absolutely, which is bad enough," Johnson lamented. "What is far worse is that it rules much of the Christian world as well." As an antidote to this poison, Johnson devised a strategy he called "the wedge":

> A log is a seeming solid object, but a wedge can eventually split it by penetrating a crack and gradually widening the split. In this case the ideology of scientific materialism is the apparently solid log. The widening crack is the important but seldom-recognized difference between the facts revealed by scientific investigation and the materialist philosophy that dominates the scientific culture.

He himself would become "the leading edge of the Wedge of Truth" and "make the initial penetration into the intellectual monopoly of scientific naturalism."[3]

Darwin on Trial sold like hotcakes. Naturally, it attracted much negative attention as well. One justifiable criticism of Johnson was that he behaved too much like a lawyer—he concentrated on the offense and paid no heed to defense. What could Johnson offer as a substitute for Darwinism? An answer came five years later in *Darwin's Black Box* (1996), by the Lehigh University biochemist Michael Behe (b.1952). In this book Behe invited his readers to focus on something he called "irreducible complexity"—a system where all of the parts are intricately matched together in such a way that the system breaks down if any part is removed. "An irreducibly complex system cannot be produced directly (that is, by continuously improving the initial function, which continues to work by the same mechanism) by slight, successive modifications of a precursor system, because any precursor to an irreducibly complex system that is missing a part is by definition nonfunctional." Behe argued, surely with some truth, that any "irreducibly complex biological system, if there is such a thing, would be a powerful challenge to Darwinian evolution. Since natural selection can only choose systems that are already working, then if a biological system cannot be produced gradually it would have to arise as an integrated unit, in one fell swoop, for natural selection to have anything to act on."[4]

But are there such systems? Behe thought that there are. He highlighted an example familiar to those who work on microorganisms. To move about, some bacteria use a cilium, a kind of whip that contains its own power unit and works by a paddling or rowing motion. Others go a different route, using a flagellum (another whiplike appendage) as a kind of propeller. "The filament [the external part] of a bacterial flagellum, unlike a cilium, contains no motor protein; if it is broken off, the filament just floats stiffly in the water. Therefore, the motor that rotates the filament-propeller must be located somewhere else. Experiments have demonstrated that it is located at the base of the flagellum, where electron microscopy shows several ring structures occur." You need a connector between the external part of the flagellum (the filament) and motor, and lo and behold nature provides it: a "hook protein" is right there to do the job as needed. The system is highly complex in itself, and added to that is the fact that the motor has its own energy source. Energy is created continuously by the cell, rather than, more conventionally, by drawing on energy stored in various complex molecules. To Behe, none of this could have evolved or appeared in a gradual way, bit by bit. It all had to be there, up and running, in one creative act, and this required conscious design and flawless execution—by an intelligent designer.[5]

Behe's empirical argument soon received theoretical support from the philosopher-mathematician William A. Dembski (b. 1960). "The key step in formulating Intelligent Design as a scientific theory," reasoned Dembski, was "to delineate a method for detecting design." According to his "three-stage Explanatory Filter," three notions underlie the concept of design: contingency, complexity, and specification. Design has to be contingent. Things that follow by blind law give no evidence of design. The ball falls and bounces. No design here. Second, design has to be complex. The number 2 followed by 3 does not catch your attention. But the succession of prime numbers up to 101 makes you think that something interesting is going on. Finally, design demands a certain independence or specification—you cannot confer the criteria for design after the fact. Drawing a bull's-eye around an arrow that has already landed is not design. But an arrow that hits a bull's-eye that was specified in advance makes you think that the arrow's position is not random.[6]

However it may be qualified, the point of the ID movement is to promote the intellectual respectability of interventions outside the natural order of things by including them under the rubric "science." We can certainly take with a pinch of salt the movement's claims about the absence of theological links and motives. Asked about how he thinks complexity first occurs, Michael Behe responded: "In a puff of smoke!" Pressed with the question, "Do you mean that the Intelligent Designer suspends the laws of physics through working a miracle?" he replied, "Yes." Dembski has stated bluntly that "a scientist in trying to understand some aspect of the world, is in the first instance concerned with that aspect as it relates to Christ—and this is true regardless of whether the scientist acknowledges Christ." He added later: "Intelligent design is just the Logos theology of John's Gospel restated in the idiom of information theory." In the language of the Good Book: "In the beginning was the word, and word was with God, and the word was God" (John 1:1).[7]

Despite these traces of God-talk, it would be a mistake simply to categorize the ID movement as creationist without qualification. Unlike young-earth creationists, the ID theorists make no appeal to Genesis and seem totally unconcerned about the antiquity of life on earth. Many if not most of the leaders subscribe to, or at least are open to, some form of evolution. Behe: "I'm an 'evolutionist' in the sense that I do think natural selection explains some things . . . But from what I see, the evidence only shows natural selection explaining rather small changes, and I see profound difficulties in thinking that it explains much more than trivial changes. It is fine by me if common descent is indeed true, and there is some sort of designed program to power changes over time (i.e., evolution). And I think things like pseudo-genes are strong arguments for common descent. So again I'm an 'evolutionist' in that sense." Dembski: "Right now I'm inclined toward a preprogrammed form of evolution in which life evolves teleologically (humanity being the end of the evolutionary process)." Even Johnson is not a flat-out, take-no-hostages opponent of evolution: "I agree . . . that breeding groups that became isolated on an island often vary from mainland species as a result of interbreeding, mutation, and selection. This is change within the limits of a pre-existing type, and not necessarily the means by which the types came into existence in the first place. At a more general level, the pattern of relationship among plants and animals suggests that they may have been produced by some process of development from some common source."[8]

Beginning with the appearance of Johnson's *Darwin on Trial* ID stirred a heated debate, which centered on the question, Is it science or, crudely put, merely "the same old creationist bullshit dressed up in new clothes"?[9] In the United States at least questions of this kind are often resolved by courts and legislatures, not by scientists and philosophers. Thus in 1981 I found myself in a Little Rock, Arkansas, courtroom testifying as an expert witness in a case to decide the constitutionality of a recently passed state law mandating "balanced treatment" for what were being called creation science and evolution science. Taking my advice, the judge decided that "the essential characteristics of science" included naturalness, tentativeness, testability, and falsifiability—and ruled that creation science failed to meet these criteria.[10] In 1987 the U.S. Supreme Court adopted this reasoning when it invalidated a similar Louisiana law. Science, according to this definition, is an attempt to understand the world according to unbroken law—by law in this context, I mean law of nature, and by law of nature I mean something that holds universally and that we expect to hold universally. Using this definition, plate tectonics qualifies as science because it is something that works according to unbroken natural law. *Paradise Lost* is not science because it involves all sorts of miraculous happenings that go on outside the realm of nature.

The insistence that science restrict itself to nature goes back at least to the early nineteenth century. In 1837, when the Cambridge philosopher-historian William Whewell (1794–1866) published his *History of the Inductive Sciences,* he felt it necessary to bring in God to explain the origin of organisms, but he carefully noted that this was not science: "[W]hen we inquire whence they came into this our world, geology is silent. The mystery of creation is not within the range of her legitimate territory; she says nothing, but she points upwards." When faced with the "mystery of mysteries," the origins of organisms, Charles Darwin did not take the Whewell cop-out option. He worked and worked until he found natural selection. When his American correspondent Asa Gray wanted to bring God back in through directed mutations, Darwin was considerably less than sympathetic.[11]

One of the most intriguing attempts to provide philosophical cover for ID's attempt to recognize supernaturalism as science has come from the respected philosopher of religion Alvin Plantinga (b. 1932), who has long harbored antipathy toward evolution. While agreeing that recourse to miracles might be a "science stopper," he nevertheless sees no reason to limit science to the natural. "Obviously we have no guarantee that God has done everything by way of employing secondary causes, or in such a way as to encourage further scientific inquiry, or for our convenience as scientists, or for the benefit of the National Science Foundation," he writes. "Clearly we cannot sensibly insist in advance that whatever we are confronted with is to be explained in terms of something *else* God did; he must have done *some* things directly. It would be worth knowing, if possible, which things he *did* do directly; to know this would be an important part of a serious and profound knowledge of the universe." He would have us use the term *Augustinian science* for regular science plus miracles.[12]

Although the scientific community simply ignored Plantinga's idiosyncratic proposal and others like it, the antievolution Kansas State Board of Education in 2005 voted no longer to limit science to the natural. Ditching its old definition of science as "the human activity of seeking natural explanations for what we observe in the world around us," the board substituted a description of science as "a systematic method of continuing investigation that uses observation, hypothesis testing, measurement, experimentation, logical argument and theory building to lead to more adequate explanations of natural phenomena." This prompted the conservative psychiatrist-columnist Charles Krauthammer (b. 1950) to explode: "In order to justify the farce that intelligent design is science, Kansas had to corrupt the very definition of science, dropping the phrase '*natural* explanations for what we observe in the world around us,' thus unmistakably implying—by fiat of definition, no less—that the supernatural is an integral part of science."[13]

That same year a group of concerned parents in the Dover (Pennsylvania) Area School District filed a lawsuit against the board's decision to recommend *Of Pandas and People* as a way of making students "aware of gaps/problems in Darwin's theory and of other theories of evolution including, but not limited to, intelligent design."[14] One of the witnesses for the defense, Steve William Fuller (b. 1959), a postmodern "social epistemologist" from England, insisted that "there is nothing especially unscientific about aiming to change the ground rules of science." He argued that ID, despite its appeal to "supernatural causation," constituted legitimate "science," and he objected to the scientific community's "dogmatic" commitment to methodological naturalism, which he denied was "a ground rule of science."[15] In the end the judge rejected such perverse reasoning, ruling that ID was "not science" because it invoked "supernatural causation" and failed "to meet the essential ground rules that limit science to testable, natural explanations."[16] The effort to impose ID on the children of Dover, he memorably declared, was a "breathtaking inanity."

THAT CREATIONISM IS A UNIQUELY

AMERICAN PHENOMENON

Ronald L. Numbers

As insidious as it [creationism] may seem, at least it's not a
worldwide movement . . . I hope everyone realizes the extent
to which this is a local, indigenous, American bizarrity.

—Stephen Jay Gould, commenting in an Associated Press
interview (2000)

[C]reationism is an American institution, and it is not only
American but specifically southern and southwestern.

—Richard C. Lewontin, Introduction to *Scientists
Confront Creationism* (1983)

Ever since the outbreak in the 1960s of so-called scientific creationism, which squeezed the history of life on earth into little
more than 6,000 years, critics have consoled themselves with the
notion that the movement could be geographically contained.
As the American paleontologist and anticreationist Stephen Jay
Gould assured New Zealanders in 1986, they had little to fear
from scientific creationism because the movement was so "peculiarly American." In his opinion, it stood little chance of "catching
on overseas." Already, however, it was spreading beyond the con-
fines of the United States. Despite its "Made in America" label,
creationism had begun to break out among conservative Protestants in various parts of the world. At first the leading carriers

were men such as Henry M. Morris and Duane Gish from the Institute for Creation Research (ICR), which published creationist tracts in some two dozen languages. After the mid-1990s leadership passed increasingly to Kenneth A. Ham at Answers in Genesis (AiG), a Kentucky-based operation located just south of Cincinnati. In less than a decade he and his AiG colleagues created a network of international branches and distributed books in languages ranging from Afrikaans and Albanian to Romanian and Russian. During that time scientific creationism and its younger cousin "intelligent design" also spread from their evangelical Protestant bases to Catholicism, Eastern Orthodoxy, Islam, Judaism, and beyond.[1]

During the century or so following the publication of Charles Darwin's *Origin of Species* (1859) most conservative Christian antievolutionists, at least those who expressed themselves on the subject, accepted the evidence of the antiquity of life on earth while rejecting the transmutation of species and any relationship between apes and humans. Only a small minority, found largely among the Seventh-day Adventist followers of the prophet Ellen G. White, insisted on the special creation of all life forms 6,000–10,000 years ago and on a universal flood at the time of Noah that buried most of the fossils. Following the publication of *The Genesis Flood* (1961) by the fundamentalists John C. Whitcomb, Jr., and Henry M. Morris, this radical alternative broke from its Adventist moorings and cut a broad swath through conservative Protestantism. About 1970 its advocates, hoping to make their biblical views palatable for public-school consumption, euphemistically began calling it "scientific creationism" or "creation science." Because of its rejection of geological ages, it also became known as young-earth creationism.[2]

Few countries outside the United States gave creation science a warmer reception than Australia. A visit from Morris in 1973 initially sparked interest in creation science. Within five years young-earth creationists, led by Ham, a schoolteacher, and Carl Wieland, a physician, had organized the Creation

Science Foundation (CSF), which from its headquarters in Brisbane quickly became the center of antievolutionism in the South Pacific. Despite its name, the CSF stressed the biblical arguments for creationism. "Unshackled by constitutional restraints" on the teaching of religion in public schools, as one critic put it, Australian creationists were "not as coy as their American counterparts in declaring their evangelical purposes." Antievolutionists in Australia celebrated in August 2005, when the minister of education, a Christian physician named Brendan Nelson, came out in favor of exposing students to both evolution and intelligent design. "As far as I'm concerned," he explained, "students can be taught and should be taught the basic science in terms of the evolution of man, but if schools also want to present students with intelligent design, I don't have any difficulty with that. It's about choice, reasonable choice."[3]

Similar developments occurred in New Zealand, only more slowly and with less fanfare. A rightward shift politically and religiously had created by the 1980s a fertile field in which creationism could grow. In 1992 New Zealand creationists set up an "NZ arm" of the CSF, called Creation Science (NZ). Three years later the *New Zealand Listener* surprised many of its readers by announcing that "God and Darwin are still battling it out in New Zealand schools." In contrast to the common image of a thoroughly secular educational system, the popular magazine revealed that "specialists with science degrees" had been promulgating creationism in the country's classrooms, where they often discovered a sympathetic audience, particularly among Maori and Pacific Islanders, who tended to view evolution with suspicion. As one Maori leader observed, "the ultimate in alienation would be to be a Maori and an EVOLUTIONIST." Against the odds—and the assurances of Gould—scientific creationism had established a beachhead deep in the Antipodes.[4]

Writing in 2000, one observer claimed that "there are possibly more creationists per capita in Canada than in any other Western country apart from the US." Though counterintuitive, the claim may have been true. In 1993 *Maclean's,* "Canada's Weekly Newsmagazine," shocked many readers when it carried a public-opinion poll showing that "even though less than a third of Canadians attend a religious service regularly . . . 53% of all adults reject the theory of scientific evolution." In 2000, at a time when some public schools in Canada were downplaying evolution, a Pentecostal lay preacher, Stockwell Day, ran unsuccessfully for prime minister while claiming "that the Earth is 6,000 years old, humans and dinosaurs roamed the planet at the same time and that Adam and Eve were real people."[5]

Before 2002 few people in Great Britain except evangelicals gave much thought to creationism. That year, however, the British press drew attention to a creationist "scandal" in Gateshead, where, as one reporter put it, "Fundamentalist Christians who do not believe in evolution have taken control of a state-funded secondary school in England." Warning of the spread of "US-Style Creationism," a broad range of concerned citizens expressed outrage that young-earth creationism had infiltrated British schools. The outspoken Oxford evolutionist Richard Dawkins condemned the teaching of creationism as an act of "educational debauchery." At the same time a group of prominent Christians blamed Dawkins himself for exacerbating the problem by "portraying science as an irreligious activity." Ham and his associates at AiG reveled in the unexpected publicity. "It's exciting," they said, "to see how God's *enemies* are bringing national attention—free of charge—to AiG's efforts to defend the authority of God's Word, and to call the languishing church in Britain back to its roots in Genesis!"[6]

By late 2005 antievolutionism in the United Kingdom had grown to such proportions that the retiring president of the Royal Society, Britain's national academy of science, devoted his farewell address to warning that "the core values of modern science are under serious threat from fundamentalism." Within months the BBC shocked the nation when it announced the results of a poll showing that "four out of 10 people in the UK think that religious alternatives to Darwin's theory of evolution should be taught as science in schools." The survey, conducted in connection with a broadcast ominously called *A War on Science*, revealed that only 48 percent of Britons believed that the theory of evolution "best described their view of the origin and development of life." Twenty-two percent said that "creationism" best described their views, 17 percent favored "intelligent design," and 13 percent remained undecided. Teachers reported strong creationist sentiment among their students. A sixth-form biology teacher in London complained that "the vast majority" of her brightest students, including many headed for careers in the health professions, rejected evolution. Often they came from Pentecostal, Baptist, or Muslim families. "It's a bit like the southern states of America," she noted.[7]

Elsewhere in western Europe creationists were making similar inroads. A poll of adult Europeans revealed that only 40 percent believed in naturalistic evolution, 21 percent in theistic evolution, and 20 percent in a recent special creation, while 19 percent remained undecided or ignorant. The highest concentrations of young-earth creationists were found in Switzerland (21.8 percent), Austria (20.4 percent), and Germany (18.1 percent). Outside the German-speaking countries the strongest support for creationism in western Europe came from the Netherlands, where creationists had been active since the 1970s and where over half the population professed to believe in God and 8 percent subscribed to the inerrancy of the Bible. In the spring of 2005 the Dutch science and education minister triggered a fierce debate in parliament by suggesting that the teaching of intelligent design might help to heal religious rifts. "What unites Muslims, Jews, and Christians is the notion that there is a creation," she said optimistically. "If we succeed in connecting scientists from different religions, it might even be applied in schools and lessons." The threat of state-supported antievolutionism prompted one science writer in Amsterdam to ask, "Is Holland Becoming the Kansas of Europe?"[8]

Italian antievolutionists formed a society in the early 1990s dedicated to introducing "into both public and private schools the biblical message of creationism and the scientific studies that confirm it," but most Italian academics ignored the threat until early in 2004, when the right-wing political party Alleanza Nazionale began dismissing evolution as a "fairytale" and linking Darwinism to Marxism. About the same time the Italian minister of education, universities, and research shocked the nation with her plan to eliminate the teaching of evolution to students ages eleven to fourteen. An estimated 46,000 to 50,000 irate Italians, including hundreds of prominent scientists, rose up to protest what they regarded as "part of a growing antiscientific trend in our country," forcing the minister to back down. To defend Darwinism against creationism and to restore "sanity" to science education, concerned Italian scientists in 2005 formed the Society for Evolutionary Biology.[9]

Almost immediately after the fall of the Berlin Wall in 1989 and the dissolution of the Soviet Union two years later, conservative Christians began to flood the formerly communist countries of eastern Europe. Within a few years creationist missionaries had successfully planted new societies in Poland, Hungary, Romania, Serbia, Russia, and the Ukraine. In 2004 the minister of education in Serbia, who favored the teaching of creationism, informed primary school teachers that they should no longer have students read a "dogmatic" chapter on Darwinism in the commonly used eighth-grade biology textbook. The following year the Romanian ministry of education granted permission for teachers in both public and Christian schools to elect to use a creationist alternative to the standard biology textbook. In 2006 Poland's minister of education repudiated evolution, while his deputy dismissed it as "a lie . . . an error we have legalized as a common truth."[10]

Creationism in Russia flourished in the early 1990s under the energetic leadership of Dmitri A. Kouznetsov, who claimed to have earned three doctorates in science. (Later he was exposed as a fraud and briefly jailed in the United States for passing bad checks.) In 1994 the department of extracurricular and alternative education of the Russian ministry of education cosponsored a creationist conference, at which a deputy minister of education urged that creationism be taught to help restore academic freedom in Russia after years of state-enforced scientific orthodoxy. "No theory," stated one academician, "should be discounted after the long Communist censure." The ministry enlisted Duane Gish from the ICR to develop curricular materials on the subject of origins. By the late 1990s reports were reaching North America "that Russian scientists desperately need resources to stem the rising tide of creationism in their country." One observer described St. Petersburg as being "flooded with Russian translations of books and pamphlets about 'creation science.' "[11]

After a very slow start in Latin America, creationists witnessed "an explosion" of interest in the late 1990s, paralleling that of evangelical Christianity generally. Nowhere in South America did antievolutionists make deeper inroads than in Brazil, where, according to a survey in 2004, 31 percent of the population believed that "the first humans were created no more than 10,000 years ago" and the overwhelming majority favored teaching creationism. In 2004 the evangelical governor of the state of Rio de Janeiro announced that public schools would be teaching creationism. "I do not believe in the evolution of species," she declared. "It's just a theory." Evolutionists tried to mount a protest, but, unlike their colleagues in Italy, they failed to generate much interest. The Catholic majority in the country, explained a dispirited scientist, had become confused and overwhelmed by the aggressive Protestants who "imported creationism from the U.S."[12]

In Asia, the Koreans emerged as *the* creationist powerhouse, propagating the message at home and abroad. Since its founding in the winter of 1980–1981, the largely Christian Korea Association of Creation Research (KACR) has flourished. Within fifteen years of its founding the association had spawned sixteen branches, recruited several hundred members with doctorates of one kind or another, and published dozens of creationist books and a bimonthly magazine, *Creation,* with a circulation of 4,000. By 2000 the membership stood at 1,365, giving Korea claim to being the creationist capital of the world, in density if not in influence. In the 1980s the association began proselytizing among Koreans on the West Coast of North America, which led to the formation of several chapters. In 2000 the KACR dispatched its first creation science missionary, to Muslim Indonesia, where the association had been sending lecturers for some time.[13]

For decades creationism remained largely confined to Christian enclaves. But in the mid-1980s the ICR received a call from the Muslim minister of education in Turkey saying that "he wanted to eliminate the secular-based, evolution-only teaching dominant in their schools and replace it with a curriculum teaching the two models, evolution and creation, fairly." Because they followed the Qur'an in believing that Allah had created the world in six days (at an unspecified time in the past), many Muslims found the arguments of Christian creationists appealing. As a result of the contact between the ministry of education and the ICR, several American creationist books were translated into Turkish, and complimentary copies of *Scientific Creationism* were sent to every science teacher in the public schools of Turkey.[14]

In 1990 a small group of well-to-do young Turks in Istanbul formed the Science Research Foundation (BAV), dedicated to promoting an immaterial cosmology and opposing evolution. At the center of the organization was the charismatic Harun Yahya, the pen name of Adnan Oktar, a student of interior design and philosophy turned imam. Assisted by his followers, he produced nearly 200 books, including *The Evolution Deceit: The Collapse of Darwinism and Its Ideological Background* (1997). Evolution, he explained, denied the existence of Allah, abolished moral values, and promoted communism and materialism. Over the next decade BAV gave away millions of copies of this book, in languages ranging from Arabic to Urdu.[15]

Until the late twentieth century even the most orthodox Jews, who typically did not accept evolution, rarely paid attention to Christian creationists—or to any kind of scientific argument against evolution. In 2000, however, a group of Jewish antievolutionists in Israel and the United States formed the Torah Science Foundation (TSF). Behind this initiative lay the influence of the late Lubavitcher rebbe Menachem Mendel Schneerson, who insisted that evolution had "not a shred of *evidence* to support it." The head of the TSF, Eliezer (Eduardo) Zeiger, a professor of ecology and evolutionary biology at the University of California, Los Angeles, advocated what he called "*Kosher* Evolution." Like Christian young-earth creationists, he accepted microevolution while rejecting macroevolution. To distinguish his views from theirs, he stressed his "access to the inner wisdom of the Torah, which includes Kabbalah and Chassidic philosophy."[16]

At first few observers noticed the spread of creationism outside the United States. In 2000, however, the British magazine *New Scientist* devoted a cover story urging readers to "Start Worrying Now" because "From Kansas to Korea, Creationism Is Flooding the Earth." Although this development seemed almost "beyond belief," the magazine described creationism as "mutating and spreading" around the world, "even linking up with like-minded people in the Muslim world." Just five years later representatives from national academies of science around the world joined in signing a statement supporting evolution and condemning the global spread of "theories not testable by science."[17] Contrary to almost all expectations, geographical, theological, and political barriers had utterly failed to contain what had begun as a distinctively American movement to create a Bible-based alternative to evolution.

THAT MODERN SCIENCE HAS SECULARIZED
WESTERN CULTURE

John Hedley Brooke

I shall suggest that the existence of God is a scientific hypothesis like any other.

—Richard Dawkins, *The God Delusion* (2006)

Belief in supernatural powers is doomed to die out, all over the world, as a result of the increasing adequacy and diffusion of scientific knowledge.

—Anthony F. C. Wallace, *Religion: An Anthropological View* (1966)

In November 2006 a new "Center for Inquiry" held its inaugural press conference in Washington, D.C. Its aim? To "promote and defend reason, science, and freedom of inquiry in all areas of human endeavor." This was deemed necessary because of "the resurgence of fundamentalist religions across the nation, and their alliance with political-ideological movements to block science." Prominent scientists signed a declaration lamenting the "persistence of paranormal and occult beliefs" and a "retreat into mysticism." Their contention was that public policies should be shaped by secular values and that science and secularism are "inextricably linked."[1] If they are, and if Richard Dawkins (b. 1941) is right to regard the existence of God as a vulnerable "scientific" hypothesis, it seems reasonable to assert that scientific progress has been the principal cause of secularization. This claim, however, belongs to a category of "obviously true" propositions that, on closer examination, turn out to be largely false.[2]

As with many myths, the proposition that "science causes secularization" contains elements of truth. Definitions of secularization usually refer to the displacement of religious authority and control by civic powers and the loss of beliefs characteristic of religious traditions. If scientific knowledge has had a corrosive effect, lending support to world views less dominated by the supernatural, one can see why correlations are made between scientific progress and secularization. If, as Thomas Hobbes (1588–1679) argued, the origins of religious belief lay in the fear and incomprehensibility of nature's forces, then, as knowledge replaced ignorance, superstition would surely recede. Where scientific explanation remained incomplete, religious thinkers might plug the gaps with their gods, but further scientific advance would then again reduce their influence, and so on. Moreover, the content of scientific theories has sometimes clashed with conventional readings of sacred texts. This was true of explanations of the earth's motion in Galileo's day and of evolutionary accounts of human origins in Darwin's. Are these not instances of science favoring secularization? The introduction of Western education, philosophy, and technology in nineteenth-century India had consequences described as a "massive and thoroughgoing secularization."[3] If further evidence were needed from within other cultural traditions, such as in Judaism, many eminent scientists of the twentieth century were "non-Jewish Jews."[4]

The question is whether these elements of truth constitute the whole truth or whether the "myth" discussed in this chapter has escaped criticism partly because of its manifest use in promoting science and suppressing religion. The "science causes secularization" formula is surely misleading. Are these two processes really linked "inextricably"? Historical evidence suggests not. An association of scientific rationality with a secular mentality

is commonly assumed by natural scientists, many of whom re-
joice in what they see as the corrosive effects on religion of rigorous empirical methods. By contrast, many social scientists
now reject what was once known as the secularization thesis—
that an inexorable depletion of religious authority and function
is irreversible in societies permeated by science and technology.[5]
Paradoxically, the urgency of the plea from the Center for Inquiry
is more supportive of a view found among social scientists—
that religious beliefs and practices may flourish and even re-
gain allegiance in scientifically and technologically advanced
societies.

In controlling natural forces, science-based technologies have
certainly far surpassed the results of contemplation or suppli-
cation. Their effects on religious practice, however, have been
strangely diverse. Indirectly, by facilitating new modes of transport and recreation, they have contributed seductive alternatives
to the religious life. But new technologies may also facilitate religious observance; for example, in some Jewish communities
pre-programmable elevators and ovens have been used to keep
Sabbath injunctions. At the level of scientific theory there is the
complication that the form and even the content of scientific theories may *reflect* the values enshrined within a particular society
as much as they may *produce* them.[6]

There is an important difference between secularization *of*
science and secularization *by* science. Religious language had
largely disappeared from technical scientific literature by the end
of the nineteenth century; but this does not mean that religious
beliefs were no longer to be found among scientists. Crucially,
the cultural significance given to scientific discoveries and theories has depended on the preconceptions of the times. Scientists
with religious convictions have often found confirmation of their
faith in the beauty and elegance of the mechanisms their research uncovers.[7] For the seventeenth-century astronomer Johannes Kepler (1571–1630), the mathematical elegance of the
laws describing planetary motion prompted his confession that

he had been carried away by "unutterable rapture at the divine spectacle of heavenly harmony."[8] A contemporary example would be the former director of the Human Genome Project, Francis Collins (b. 1950), who sees his work as the unraveling of a God-given code.[9]

Instead of regarding science as the principal agent of secularization, it is more accurate to say that scientific theories have usually been susceptible to both theistic and naturalistic readings. Historically they have provided resources for both.[10] Sometimes the same concept has been manipulated to generate a sense of the sacred and of the profane. This is true of what is arguably the most corrosive of all scientific theories: Charles Darwin's (1809–1882) theory of evolution by natural selection. For Richard Dawkins, Darwin first made it possible to be an intellectually fulfilled atheist. But it is easy to forget that among Darwin's earliest sympathizers in Britain were Christian clergymen such as Charles Kingsley (1819–1875) and Frederick Temple (1821–1902). Kingsley delighted Darwin by suggesting that a God who could "make things make themselves" was more admirable than one who simply made things.[11] Temple, who welcomed the extension of natural law because it gave analogical support for belief in moral law, was later to become archbishop of Canterbury, primate of the English church.[12] Although an agnostic late in life, Darwin denied he had ever been an atheist and frequently referred to evolutionary outcomes as the result of laws impressed on the world by a creator.[13]

Instead of seeing science as intrinsically and inextricably secular, it is more correct to see it as neutral with respect to questions concerning God's existence. Interestingly, this is how it *was* seen by Darwin's most vigorous popularizer, Thomas Henry Huxley (1825–1895), from whom Dawkins has therefore to distance himself.[14] For Huxley, science was neither Christian nor anti-Christian but extra-Christian, meaning that it had a scope and autonomy independent of religious interests. Hence his insistence that Darwin's theory had no more to do with theism than did the first book of Euclid, meaning it had no bearing on the deeper question, whether evolutionary processes themselves might have been seeded in an original design.[15] The pressing question is not whether Darwin's theory has been used to justify unbelief—it has, many times—but whether its use as justification conceals other, more important reasons for unbelief. As James R. Moore makes clear in Myth 16, the main reasons Darwin gave for his unbelief derived not from the role he gave natural causes in explaining the origin of species. Like other Victorian thinkers, Darwin reacted strongly against evangelical Christian preaching on heaven and hell. Members of his family had been freethinkers: his grandfather Erasmus had been an early advocate of organic evolution, his father was probably an atheist, his brother Erasmus certainly so. The doctrine that after death they would suffer eternal damnation was, for Charles, a "damnable doctrine."[16] He was also sensitive to the extent of pain and suffering, which he described as one of the strongest arguments against belief in a beneficent deity. Each of these concerns was crystallized by deaths in his family—that of his father in the late 1840s and of his ten-year-old daughter Annie early in 1851. Darwin did believe that, as science advanced, appeals to the miraculous became more incredible; but his loss of faith had deeper existential roots.[17] The myth consists in the view that science, *more than any other factor,* is the agent of secularization.

Can the factors really be weighed? Impressive attempts to do so seem to confirm that to give primacy to science is a mistake. Years ago a sociologist investigated the reasons given by secularists, in the period from 1850 to 1960, for their conversion from Christianity to unbelief.[18] Reading the direct testimony of one hundred fifty unbelievers and related evidence from two hundred additional biographies, she found that science barely featured at all. Conversions to unbelief were often associated with a change from conservative to more radical politics, with religion being rejected as part of established, privileged society. The reading of radical texts, such as *The Age of Reason* (1794–1795) by Thomas Paine (1737–1809), was another prominent influence.[19] Ironically, one of the other frequently mentioned books was the Bible itself, close study of which revealed what were seen as inconsistencies, absurdities, or (particularly in the Old Testament) depictions of a vengeful and anthropomorphic deity. In 1912 the president of the National Secular Society in Britain insisted that biblical stories of "lust, adultery, incest and unnatural vice" were "enough to raise blushes in a brothel."[20] The fact that every Christian sect, indeed every religion, claimed its own hotline to the truth was a prevalent consideration having nothing to do with science. The perceived immorality of religious doctrines, particularly those concerning an afterlife, and the perceived immoral behavior of some priests fueled a rejection of religious authority. The argument that atheists could be as morally upright as believers also took its toll. Historical more than scientific research was proving subversive as biblical writers came to be seen not as timeless authorities but as unreliable products of their own culture.

A distinguished anthropologist, Mary Douglas, has observed that those who imagine science the principal cause of secularization forget that religious activity is grounded in social relations, not primarily in concepts of nature.[21] Consequently it is wiser to look to long-term changes in social structure and to changes in religion itself if one wishes to understand the momentum of secularity. In the mid-nineteenth century, when the idea of conflict between science and religion first caught the public eye, the changes that most precipitated secular reaction came from *within* both Protestant and Catholic Christianity. Claims for the inerrancy of Scripture could lead to an unattractive bibliolatry. Similarly, claims for papal infallibility on matters of faith and doctrine (1870) and the strictures of the Syllabus of Errors published by the Catholic church in 1864 antagonized many, including John William Draper, whose *History of the Conflict between Religion and Science* (1874) promulgated a resilient, if overstated, thesis that science and Catholic Christianity were mortal enemies.[22] Earlier, the intolerance shown by religious denominations to dissenters had evoked critical attitudes during the Enlightenment, especially in France, where Voltaire's quarrel with the Catholic church was primarily on moral and political grounds. As Voltaire (1694–1778) realized, Newton's science actually supported theism rather than atheism—and continued to do so.[23]

In modern times, the expansion of secularism can be correlated with social, political, and economic transformations having little direct connection with science. Historians point to increases in social and geographical mobility that have fractured communities once bound by common religious values. The growth of capitalism, commerce, and consumerism has fostered a pervasive hedonism that threatens commitment to religious institutions and their long-term goals. Competing attractions have encouraged the marginalizing of religious worship. Secular values have been heavily promoted in the sphere of education and by the media. In some countries religious solidarity has been displaced by national solidarity or by the ideology of political parties. That such transformations have taken place at different rates and to different degrees in different cultures means there is "no consistent relation between the degree of scientific advance and a reduced profile of religious influence, belief and practice."[24]

Because different countries and cultures have experienced the tension between secular and religious values in contrasting ways, there is no one, universal process of secularization that can be ascribed to science or to any other factor. The freedom in the United States to believe more or less anything from a smorgasbord of ideas, ideals, and therapies contrasts sharply with the repressive constraints at work in societies such as the former East Germany, where, under a communist regime, such freedom of expression was denied. Interestingly, where nations with a long

religious tradition have been oppressed by a foreign power, religion has often reinforced a sense of national identity that breaks out of its chains with a new vitality once freedom has been gained. The strength of Catholicism in Poland is a modern example. The collapse of communist ideology within Russia itself allowed an old union of faith and nation to be reignited. A history of secularization in France would be very different from its history in the United States, where centralizing tendencies of all kinds have been resisted. In another twist to the story, aggressive antireligious remarks made by vociferous scientists frequently elicit strong reactions from those who find in their religious practice a meaning and orientation that scientific knowledge alone seems unable to provide.

Undoubtedly science can be connected to secularization by definition. The *Shorter Oxford English Dictionary* defines *secular* to mean "the world," especially in contrast to the church. The connection with science then becomes almost necessary by definition because scientists do devote themselves to the study of the world, its history, and its mechanisms. But this oversimplifies a complex question. Many scientists have found it possible to harmonize their faith with their science. Predictions made long ago that future scientific progress would banish religion now seem naïve. Just before World War I the psychologist James Leuba (1867–1946) conducted a survey in which a thousand American scientists were asked whether they believed in a personal God "to whom one may pray in expectation of receiving an answer." The percentage subscribing to that belief was 41.8. Leuba predicted that with the advance of science the proportion would shrink.[25] When the results of an identical survey were reported in *Nature* in 1998, the percentage was almost identical: 39.3.[26] There is evidence of a higher degree of religious skepticism among the most eminent scientists, and there is a perception that there are relatively few theists among biologists.[27] But the data reported in *Nature* at least qualify the claim that science necessarily leads to secularization. This claim is a myth. The news that it is so may be unwelcome to scientists wishing to invest their science with excessive cultural significance. As a leading expert on secularization has recently noted, it is an engaging oddity that individual scientists do not believe the data about what they believe.[28]

NOTES

INTRODUCTION

Epigraphs: Andrew Dickson White, "The Battle-Fields of Science," *New-York Daily Tribune,* 18 December 18, 1869, p. 4. John William Draper, *History of the Conflict between Religion and Science* (New York: D. Appleton, 1874), vi.

1. White, "Battle-Fields of Science," 4; Charles Kendall Adams, "Mr. White's 'Warfare of Science with Theology,' " *Forum* (September 1896): 65–78, at 67 (outcry); Elizabeth Cady Stanton, "Reading the Bible in the Public Schools," *The Arena* 17 (1897): 1033–37, at 1034. From time to time White wrote "New Chapters" for *Popular Science Monthly,* and in 1876 he brought out a small book, *The Warfare of Science* (New York: D. Appleton, 1876). Two decades later he published *A History of the Warfare of Science with Theology in Christendom,* 2 vols. (New York: D. Appleton, 1896). White was talked into substituting "dogmatic theology" for "religion" by his research associate; see Henry Guerlac, "George Lincoln Burr," *Isis* 34 (1944): 147–52. The best history of the conflict thesis remains James R. Moore, *The Post-Darwinian Controversies: A Study of the Protestant Struggle to Come to Terms with Darwin in Great Britain and America, 1870–1900* (Cambridge: Cambridge University Press, 1979), part 1, 17–122; but see also David C. Lindberg and Ronald L. Numbers, "Beyond War and Peace: A Reappraisal of the Encounter between Science and Religion," *Church History* 55 (1986): 338–54.

2. Draper, *History of the Conflict,* x–xi (Catholicism), 225–26 (infallibility); Donald Fleming, *John William Draper and the Religion of Science* (Philadelphia: University of Pennsylvania Press, 1950), 31 (Eliz-

abeth), 129 (fagot); review of *History of the Conflict between Religion and Science*, by John William Draper, in *Brownson's Quarterly Review*, last ser. 3 (1875): 145 (lies); review of *History of the Conflict between Religion and Science*, by John William Draper, in *Brownson's Quarterly Review*, last ser. 3 (1875): 153–73, at 169 (murders).

3. See Ronald L. Numbers, "Aggressors, Victims, and Peacemakers: Historical Actors in the Drama of Science and Religion," in *The Science and Religion Debate: Why Does It Continue?* (New Haven, Conn.: Yale University Press, in press).

4. Peter Harrison, " 'Science' and 'Religion': Constructing the Boundaries," *Journal of Religion* 86 (2006): 81–106. See also James Moore, "Religion and Science," in *The Cambridge History of Science*, vol. 6, ed. Peter Bowler and John Pickstone (Cambridge: Cambridge University Press, in press); and Jon H. Roberts, "Science and Religion," in *Wrestling with Nature: From Omens to Science*, ed. Peter Harrison, Ronald L. Numbers, and Michael H. Shank (Chicago: University of Chicago Press, in press).

5. Thomas Cooper to Benjamin Silliman, 17 December 1833, quoted in Nathan Reingold, ed., *The Papers of Joseph Henry*, vol. 2 (Washington, D.C.: Smithsonian Institution Press, 1975), 136; "Science and Religion," *Boston Cultivator* 7 (1845): 344 (new conquest); Gardiner Spring to Benjamin Silliman, n.d., quoted in Francis C. Haber, *The Age of the World: Moses to Darwin* (Baltimore: Johns Hopkins Press, 1959), 260–63.

6. Ronald L. Numbers, "Charles Hodge and the Beauties and Deformities of Science," in *Charles Hodge Revisited: A Critical Appraisal of His Life and Work,* ed. John W. Stewart and James H. Moorhead (Grand Rapids, Mich.: Eerdmans, 2002), 77–101, from which this account is extracted. This article is reprinted as chapter 5 in Ronald L. Numbers, *Science and Christianity in Pulpit and Pew* (New York: Oxford University Press, 2007).

7. [George Frederick Holmes], "Philosophy and Faith," *Methodist Quarterly Review* 3 (1851): 185–218, at 186; James A. Secord, *Victorian Sensation: The Extraordinary Publication, Reception, and Secret Authorship of* Vestiges of the Natural History of Creation (Chicago: University of Chicago Press, 2000), 522 (multitude).

8. Charles Darwin, *The Descent of Man, and Selection in Relation to Sex,* 2 vols. (New York: D. Appleton, 1871), 1:147 (overthrow), 2:372 (quadruped); Ronald L. Numbers, *Darwinism Comes to America* (Cambridge, Mass.: Harvard University Press, 1998), 31 (earthquake); John Tyndall, "The Belfast Address," in *Fragments of Science,* 6th ed. (New York: D. Appleton, 1889), 472–534, at 530; Andrew Dickson

White, *The Warfare of Science,* with a prefatory note by Professor Tyndall (London: H. S. King, 1876). For context, see Frank M. Turner, "The Victorian Conflict between Science and Religion: A Professional Dimension," *Isis* 69 (1978): 356–76.

9. See, e.g., Moore, *The Post-Darwinian Controversies;* David C. Lindberg and Ronald L. Numbers, eds., *God and Nature: A History of the Encounter between Christianity and Science* (Berkeley: University of California Press, 1986); John Hedley Brooke, *Science and Religion: Some Historical Perspectives* (Cambridge: Cambridge University Press, 1991); David C. Lindberg and Ronald L. Numbers, eds., *When Science and Christianity Meet* (Chicago: University of Chicago Press, 2003).

MYTH 1: THAT THE RISE OF CHRISTIANITY WAS RESPONSIBLE FOR THE DEMISE OF ANCIENT SCIENCE

This chapter borrows from two previous chapters of mine: "Early Christian Attitudes toward Nature," in *Science and Religion: A Historical Introduction,* ed. Gary B. Ferngren (Baltimore: Johns Hopkins University Press, 2002), 47–56; and "The Medieval Church Encounters the Classical Tradition: Saint Augustine, Roger Bacon, and the Handmaiden Metaphor," in *When Science and Christianity Meet,* ed. David C. Lindberg and Ronald L. Numbers (Chicago: University of Chicago Press, 2003), 7–32.

Epigraph: Charles Freeman, *The Closing of the Western Mind: The Rise of Faith and the Fall of Reason* (Knopf, 2003), xviii–xix.

1. Maria Dzielska, *Hypatia of Alexandria,* trans. F. Lyra (Cambridge, Mass.: Harvard University Press, 1995), 2 (Toland), 11 (death-blow), 19 (Gibbon), 25 (Van der Waerden), 26 (Bernal). I am grateful to Ron Numbers for his help with the Hypatia material.

2. Dzielska, *Hypatia,* passim.

3. Colossians 2:8, *New English Bible,* with substitution of an alternative translation; 1 Corinthians 3:18–19, *New English Bible.*

4. Tertullian, in *The Ante-Nicene Fathers,* ed. Alexander Roberts and James Donaldson; rev. A. Cleveland Coxe (Grand Rapids, Mich.: Eerdmans, 1986), 246b.

5. Ibid., 133a.

6. Timothy David Barnes, *Tertullian: A Historical and Literary Study,* rev. ed. (Oxford: Clarendon Press, 1985), 196.

7. Emmanuel Amand de Mendieta, "The Official Attitude of Basil of Caesarea as a Christian Bishop towards Greek Philosophy and Science," in *The Orthodox Churches and the West,* ed. Derek Baker (Oxford: Blackwell, 1976), 38, 31, and 37.

8. Basil, *Homilies on the Hexameron*, in *A Select Library of Nicene and Post-Nicene Fathers of the Christian Church*, ser. 2, ed. Philip Schaff and Henry Wace, 14 vols. (New York: Christian Literature Co., 1890–1900), 8:54.

9. Ibid., 8:70.

10. Augustine, *Confessions and Enchiridion*, trans. Albert C. Outler (Philadelphia: Westminster, 1955), 341–42.

11. Augustine, *On Christian Doctrine*, trans. D. W. Robertson, Jr. (Indianapolis: Bobbs-Merrill, 1958), 65–66.

12. Augustine, *Confessions*, trans. F. J. Sheed (New York: Sheed and Ward, 1942), 201, slightly edited.

13. Augustine, *On Christian Doctrine*, 74.

14. Augustine, *Literal Meaning of Genesis*, trans. John Hammond Taylor, S.J., in *Ancient Christian Writers: The Works of the Fathers in Translation*, ed. Johannes Quasten, W. J. Burghardt, and T. C. Lawler, vols. 41–42 (New York: Newman, 1982), 42–43.

15. Augustine, *On Christian Doctrine*, 75.

16. On Bacon, see David C. Lindberg, "Science as Handmaiden: Roger Bacon and the Patristic Tradition," *Isis* 78 (1987): 518–36.

MYTH 2: THAT THE MEDIEVAL CHRISTIAN CHURCH SUPPRESSED THE GROWTH OF SCIENCE

Epigraph: John William Draper, *History of the Conflict between Religion and Science* (New York: D. Appleton, 1874), 52.

1. Robert Wilson, *Astronomy through the Ages: The Story of the Human Attempt to Understand the Universe* (Princeton, N.J.: Princeton University Press, 1997), 45.

2. Carl Sagan, *Cosmos* (New York: Random House, 1980), 335.

3. Roger Bacon, *Compendium of the Study of Theology*, ed. Thomas Maloney (Leiden: Brill, 1988), 8, which refers to the literature.

4. John Heilbron, *The Sun in the Church: Cathedrals as Solar Observatories* (Cambridge, Mass.: Harvard University Press, 1999), 3.

5. Edward Grant, "Science in the Medieval University," in *Rebirth, Reform, and Resilience: Universities in Transition, 1300–1700*, ed. James M. Kittelson and Pamela J. Transue (Columbus: Ohio State University Press, 1984), 68–102.

6. Rainer Schwinges, *Deutsche Universitätsbesucher im 14. und 15. Jahrhundert: Studien zur Sozialgeschichte des alten Reiches* (Stuttgart: Steiner Verlag, 1986), 487–88.

7. W. R. Laird, "Robert Grosseteste on the Subalternate Sciences," *Traditio* 43 (1987): 147–69, esp. 150.

8. Jan Aertsen, "Mittelalterliche Philosophie: ein unmögliches Projekt? Zur Wende des Philosophieverständnisses im 13. Jahrhundert," in *Geistesleben im 13. Jahrhundert,* ed. Jan A. Aertsen and Andreas Speer (Berlin: Walter De Gruyter, 2000), 12–28, esp. 20–21.

9. Ernest A. Moody, *The Logic of William of Ockham* (1935; New York: Russell and Russell, 1965), 211.

10. Lynn Thorndike, *University Records and Life in the Middle Ages* (New York: Columbia University Press, 1944), 26–28.

11. Ibid., 34.

12. Jacques Verger, "A propos de la naissance de l'université de Paris: contexte social, enjeu politique, portée intellectuelle," in *Schulen und Studium im sozialen Wandel des hohen und späten Mittelalter,* ed. Johannes Fried (Sigmaringen: Jan Thorbeke Verlag, 1986), 69–96, esp. 83, 94–95.

13. Edward Grant, ed., *A Source Book in Medieval Science* (Cambridge, Mass.: Harvard University Press, 1974), 50.

MYTH 3: THAT MEDIEVAL CHRISTIANS TAUGHT THAT THE EARTH WAS FLAT

Epigraphs: John William Draper, *History of the Conflict between Religion and Science* (New York: D. Appleton, 1874), 157–59. Boise Penrose, *Travel and Discovery in the Renaissance* (Cambridge, Mass.: Harvard University Press, 1955), 7. Daniel J. Boorstin, *The Discoverers: A History of Man's Search to Know Himself and His World* (New York: Random House, 1983), x.

1. For an earlier discussion of this myth, see Lesley B. Cormack, "Flat Earth or Round Sphere: Misconceptions of the Shape of the Earth and the Fifteenth-Century Transformation of the World," *Ecumene* 1 (1994): 363–85.

2. Christine Garwood, *Flat Earth: The History of an Infamous Idea* (London: Macmillan, 2007), discusses some of this controversy, focusing on the "flat-earthers" of the nineteenth century.

3. Sadly, this continues to be repeated by some textbook writers to this day, for example, Mounir A. Farah and Andrea Berens Karls, *World History: The Human Experience* (Lake Forest, Ill.: Glencoe/McGraw-Hill, 1999), and Charles R. Coble et al., *Earth Science* (Englewood Cliffs, N.J.: Prentice Hall, 1992), both intended for secondary school audiences.

4. Washington Irving, *The Life and Voyages of Christopher Columbus: Together with the Voyages of His Companions* (London: John Murray, 1828), esp. 88.

5. Jeffrey Burton Russell, *Inventing the Flat Earth: Columbus and Modern Historians* (New York: Praeger, 1991), 24; Penrose, *Travel and Discovery in the Renaissance*, 7.

6. Charles W. Jones, "The Flat Earth," *Thought* 9 (1934): 296–307, which discusses Augustine, Jerome, Ambrose, and Lactantius.

7. Thomas Aquinas, *Summa theologica*, par. I, qu. 47, art. 3, 1.3; Albertus Magnus, *Liber cosmographicus de natura locoum* (1260). See also John Scottus, *De divisione naturae*, 3.32–33.

8. Walter Oakeshott, "Some Classical and Medieval Ideas in Renaissance Cosmography," in *Fritz Saxl, 1890–1948: A Volume of Memorial Essays from His Friends in England*, ed. D. J. Gordon (London: Thomas Nelson, 1957), 245–60, at 251. For d'Ailly, see Arthur Percival Newton, ed., *Travel and Travellers in the Middle Ages* (London: Routledge and Kegan Paul, 1949), 14.

9. Isidore of Seville, *De natura rerum* 10, *Etymologiae* III 47.

10. Wesley M. Stevens, "The Figure of the Earth in Isidore's 'De natura rerum,' " *Isis* 71 (1980): 273. Charles W. Jones, *Bedae opera de temporibus* (Cambridge, Mass.: Medieval Academy of America, 1943), 367. See also David Woodward, "Medieval *mappaemundi,*" in *The History of Cartography*, ed. J. B. Harley and David Woodward, vol. 1: *Cartography in Prehistoric, Ancient, and Medieval Europe and the Mediterranean* (Chicago: University of Chicago Press, 1987), 320–21.

11. Jean de Mandeville, *Mandeville's Travels*, trans. Malcolm Letts, 2 vols. (London: Hakluyt Society, 1953), 1:129.

12. Dante, *Paradiso,* Canto 9, 84; *Inferno,* Canto 26; Geoffrey Chaucer, "The Canterbury Tales," in *The Works of Geoffrey Chaucer,* ed. F. N. Robinson (Boston: Houghton Mifflin Co., 1961), 140, line 1228.

13. Most surveys of medieval science do not mention geography. David C. Lindberg, *The Beginnings of Western Science* (Chicago: Chicago University Press, 1992), 58, devotes one paragraph to a spherical earth. J. L. E. Dreyer, *History of the Planetary Systems* (Cambridge: Cambridge University Press, 1906), 214–19, stresses Cosmas's importance, as do John H. Randall, Jr., *The Making of the Modern Mind: A Survey of the Intellectual Background of the Present Age* (Boston: Houghton Mifflin, 1926), 23, and Penrose, *Travel,* who adds the caveat that "[i]t is only fair to state that not all writers of the Dark Ages were as blind as Cosmas," 7. Jones, "Flat Earth," demonstrates the marginality of Cosmas, 305.

14. Fernando Colon, *The Life of the Admiral Christopher Columbus by His Son Ferdinand,* trans. and annotated by Benjamin Keen (West-

port, Conn.: Greenwood Press, 1959), 39; Bartolemé de las Casas, *History of the Indies*, trans. and ed. Andrée Collard (New York: Harper and Row, 1971), 27–28.

15. Richard Eden, *The Decades of the Newe Worlde or West India . . . Wrytten in Latine Tounge by Peter Martyr of Angleria* (London, 1555), 64.

16. Concerning the long voyage, see entry for 10 October 1492 in *The Diario of Christopher Columbus's First Voyage to America 1492–93*, abstracted by Fray Bartolomé de Las Casas, transcribed and translated by Oliver Dunn and James E. Kelley, Jr. (Norman: University of Oklahoma Press, 1989), 57. Concerning the prevailing wind, see Eden, *Decades*, 66.

MYTH 4: THAT MEDIEVAL ISLAMIC CULTURE WAS INHOSPITABLE TO SCIENCE

Epigraphs: Ignaz Goldziher, "Stellung der alten islamischen Orthodoxie zu den antiken Wissenschaften," *Abhandlungen der Königlich Preussischen Akademie der Wissenschaft* 8 (1916): 3–46. Rodney Stark, *For the Glory of God: How Monotheism Led to Reformation, Science, Witch-Hunts, and the End of Slavery* (Princeton, N.J.: Princeton University Press, 2003), 155. Stark recycles this observation in *The Victory of Reason: How Christianity Led to Freedom, Capitalism, and Western Success* (New York: Random House, 2005), 20–23. Steven Weinberg, "A Deadly Certitude," *Times Literary Supplement,* 17 April 2007. At the time he made this statement, Weinberg, a recipient of the Nobel Prize in Physics, was professor of physics and astronomy at the University of Texas.

1. James E. McClellan III and Harold Dorn, *Science and Technology in World History: An Introduction* (Baltimore: Johns Hopkins University Press, 1999), 203.

2. Dimitri Gutas, *Greek Thought, Arabic Culture: The Graeco-Arabic Translation Movement in Baghdad and Early 'Abbāsid Society (2nd–4th/8th–10th centuries)* (London: Routledge, 1998), 192. See also the introduction to Jan P. Hogendijk and Abdelhamid I. Sabra, eds., *The Enterprise of Science in Islam: New Perspectives* (Cambridge, Mass.: MIT Press, 2003).

3. Charles Burnett, "Arabic into Latin: The Reception of Arabic Philosophy into Western Europe," in *The Cambridge Companion to Arabic Philosophy,* ed. Peter Adamson and Richard Taylor (Cambridge: Cambridge University Press, 2005), 370–404. See also Gutas,

Greek Thought, Arabic Culture; Roshdi Rashed, *Optique et Mathématiques* (Aldershot: Variorum, 1992), especially the chapter "Problems of the Transmission of Greek Scientific Thought into Arabic: Examples from Mathematics and Optics," 199–209; Roshdi Rashed, "Science as a Western Phenomenon," in *Encyclopedia of the History of Science, Technology, and Medicine in Non-Western Cultures*, ed. Helaine Selin (Dordrecht: Kluwer Academic Publishers, 1998); A. I. Sabra, "The Appropriation and Subsequent Naturalization of Greek Science in Medieval Islam: A Preliminary Statement," *History of Science* 25 (1987): 223–43; and A. I. Sabra, "Situating Arabic Science: Locality versus Essence," *Isis* 87 (1996): 654–70.

4. Sabra, "Appropriation," 228. Emphasis added.

5. Gutas, *Greek Thought, Arabic Culture*, 2, 5, 8.

6. Quoted in David C. Lindberg, *The Beginnings of Western Science: The European Scientific Tradition in Philosophical, Religious, and Institutional Context, 600 B.C. to A.D. 1450* (Chicago: University of Chicago Press, 1992), 175.

7. Sabra, "Appropriation," 225; Sabra, "Situating Arabic Science," 658; Rashed, "Problems of Transmission," 199–200, 202–3.

8. McClellan and Dorn, *Science and Technology in World History*, 105 (recent assessment); A. I. Sabra, "Ibn al-Haytham's Revolutionary Project in Optics: The Achievement and the Obstacle," in *Enterprise of Science in Islam*, 85–118, quotation on 86. On astronomy, see George Saliba, *A History of Arabic Astronomy: Planetary Theories during the Golden Age of Islam* (New York: New York University Press, 1994); and Edward S. Kennedy, *Astronomy and Astrology in the Medieval Islamic World* (Aldershot: Ashgate, 1998).

9. A. I. Sabra, "An Eleventh-Century Refutation of Ptolemy's Planetary Theory," in *Science and History: Studies in Honor of Edward Rosen* (*Studia Copernicana* 16), ed. Erna Hilfstein and Others (Wroclaw [Breslau]: Ossolineum, 1978).

10. Albert Z. Islandar, "Ibn al-Nafis," *Dictionary of Scientific Biography*, 9:602–6. See also Nahyan A. G. Fancy, *Pulmonary Transit and Bodily Resurrection: The Interaction of Medicine, Philosophy and Religion in the Works of Ibn al Nafis (d. 1288)*, Ph.D. dissertation, University of Notre Dame, 2006. On Avicenna's influence in Europe, see Nancy G. Siraisi, *Avicenna in Renaissance Italy: The Canon and Medical Teaching in Italian Universities after 1500* (Princeton, N.J.: Princeton University Press, 1987).

11. David C. Lindberg, *Theories of Vision from al-Kindi to Kepler* (Chicago: University of Chicago Press, 1976), 58; Rashed, "Science

as a Western Phenomenon," 887–89; A. I. Sabra, "The Physical and the Mathematical in Ibn al-Haytham's Theory of Light and Vision," in *The Commemoration Volume of Bīrūnī International Congress in Tehran*, Publication 38 (Tehran: High Council of Culture and Arts, 1976); A. I. Sabra, "Some Ideas of Scientific Advancement in Medieval Islam," paper read at the annual meeting of the History of Science Society, Raleigh, N.C., 30 October–1 November 1987; Richard Power, "Eyes Wide Open," *New York Times Magazine*, 18 April 1999, 80–83.

12. Emily Savage-Smith, "Attitudes toward Dissection in Medieval Islam," *Journal of the History of Medicine and Allied Sciences* 50 (1995): 94–97.

13. Gutas, *Greek Thought, Arabic Culture*, 168.

14. Sabra, "Situating Arabic Science."

15. George Saliba, *Islamic Science and the Making of the European Renaissance* (Cambridge, Mass.: MIT Press, 2007), 21, 233–37. On the decline of Islamic science, see, e.g., Lindberg, *Beginnings of Western Science*, 180–81.

16. Saliba, *Islamic Science*, 243.

MYTH 5: THAT THE MEDIEVAL CHURCH PROHIBITED HUMAN DISSECTION

Epigraphs: Andrew Dickson White, *A History of the Warfare of Science with Theology in Christendom*, 2 vols. (New York: D. Appleton, 1896), 2:50. Official website of Senator Specter, http://specter.senate .gov/public/index.cfm?FuseAction=NewsRoom.ArlenSpecterSpeaks& ContentRecord_id=de37ab3f-a443-472b-adb7-7218fbd27df8& Region_id=&Issue_id= (accessed 28 June 2008). Specter clearly intended to refer to Pope Boniface VIII.

1. White, *History of the Warfare*, 2:31–32.

2. Katharine Park, "The Criminal and the Saintly Body: Autopsy and Dissection in Renaissance Italy," *Renaissance Quarterly* 47 (1994): 1–33; Katharine Park, *Secrets of Women: Gender, Generation, and the Origins of Human Dissection* (New York: Zone Books, 2006), 14–25.

3. Heinrich von Staden, "The Discovery of the Body: Human Dissection and Its Cultural Contexts in Ancient Greece," *Yale Journal of Biology and Medicine* 65 (1992): 223–41; Emilie Savage-Smith, "Attitudes toward Dissection in Medieval Islam," *Journal of the History of Medicine and Allied Sciences* 50 (1995): 67–110; Vivian Nutton and Christine Nutton, "The Archer of Meudon: A Curious Absence of Continuity in

the History of Medicine," *Journal of the History of Medicine and Allied Sciences* 58 (2003): 404–5, n. 10.

4. Frederick S. Paxton, *Christianizing Death: The Creation of a Ritual Process in Early Medieval Europe* (Ithaca, N.Y.: Cornell University Press, 1990), 25–27.

5. Augustine, *Confessions* 10.35, trans. Henry Chadwick (London: Oxford University Press, 1991), 211. For an excellent general history of medical learning in the period covered by this chapter, see Nancy G. Siraisi, *Medieval and Early Renaissance Medicine: An Introduction to Knowledge and Practice* (Chicago: University of Chicago Press, 1990).

6. Mondino de' Liuzzi, *Anatomy,* trans. Charles Singer, in *The Fasciculo di medicina, Venice, 1493,* ed. Charles Singer (Florence: R. Lier, 1925); see also Nancy G. Siraisi, *Taddeo Alderotti and His Pupils: Two Generations of Italian Medical Learning* (Princeton, N.J.: Princeton University Press, 1981), 110–13, and R. W. French, *Dissection and Vivisection in the European Renaissance* (Aldershot: Ashgate, 1999), chap. 2. The latter is the most reliable general history of anatomy and dissection in this period, although even it retains traces of the myth concerning the medieval prohibition.

7. Charles H. Talbot, *Medicine in Medieval England* (London: Oldbourne, 1967), 55; Darrel W. Amundsen, "Medieval Canon Law on Medical and Surgical Practice by the Clergy," *Bulletin of the History of Medicine* 52 (1978): 22–44.

8. Elizabeth A. R. Brown, "Death and the Human Body in the Later Middle Ages: The Legislation of Boniface VIII on the Division of the Corpse," *Viator* 12 (1981): 221–70; Mary Niven Alston, "The Attitude of the Church towards Dissection before 1500," *Bulletin of the History of Medicine* 16 (1944): 225–29.

9. C. D. O'Malley, *Andreas Vesalius of Brussels, 1514–1564* (Berkeley: University of California Press, 1964), 304–6.

10. Alston, "Attitude of the Church," 233–35.

11. Park, "Criminal and the Saintly Body," 7–8.

12. Ibid., 12; Martin Kemp, *Leonardo da Vinci: The Marvellous Works of Nature and Man* (Cambridge, Mass.: Harvard University Press, 1981), 257.

MYTH 6: THAT COPERNICANISM DEMOTED HUMANS FROM THE CENTER OF THE COSMOS

Epigraphs: Martin Rees, *Before the Beginning* (Reading, Mass.: Addison-Wesley, 1998), 100. Rees is Astronomer Royal of England.

"Copernican system," *Encyclopedia Britannica Online*, at http://concise
.britannica.com/ebc/article-9361576/Copernican-system (accessed 11
June 2008).

1. Sigmund Freud, *Introductory Lectures on Psycho-Analysis: A
Course of Twenty-Eight Lectures Delivered at the University of Vienna,*
trans. Joan Riviere (London: George Allen and Unwin, 1922), 240–41.
For similar statements, see, e.g., Carl Sagan, *Pale Blue Dot* (New York:
Random House, 1994), 26; and Robert L. Jaffe, "As Time Goes By,"
Natural History Magazine 115 (October 2006): 16–24. One can find
numerous other examples by performing a web search of *Copernicus*
AND *Dethrone*.

2. Much of the material in this chapter was first published in my
longer article titled "The Great Copernican Cliché," *American Journal
of Physics* 69 (October 2001): 1029–35, and is used by kind permission
of that journal.

3. Aristotle, *Physics,* bk. 4, 208b; in *The Works of Aristotle,* ed.
W. D. Ross (Oxford: Clarendon Press, 1930), vol. 2.

4. Moses Maimonides, *The Guide for the Perplexed,* trans. M.
Friedländer, 2d ed. (New York: Dutton, 1919), 118–19; Thomas Aquinas,
Commentary on Aristotle's De Caelo (1272 s.), II, xiii, 1 & xx, n. 7, in vol.
3, 202b of the Leonina ed.; trans. and quoted by Rémi Brague, "Geocentrism as a Humiliation for Man," *Medieval Encounters* 3 (1997): 187–210
(202) (who provides numerous other examples).

5. Giovanni Pico, *Oration on the Dignity of Man,* in *The Renaissance Philosophy of Man,* ed. Ernst Cassirer et al. (Chicago: University
of Chicago Press, 1948), 224; Montaigne, *An Apology of Raymond Se-
bond* (1568), in *The Essays of Michel de Montaigne,* trans. Charles
Cotton (London: George Bell, 1892), 2:134.

6. Copernicus, *On the Revolutions,* Preface; see also John Calvin,
Commentary on Genesis; both cited in *The Book of the Cosmos: Imag-
ining the Universe from Heraclitus to Hawking,* ed. Dennis Danielson
(Cambridge, Mass.: Perseus Publishing, 2000), 107, 123–24.

7. Giovanni Maria Tolosani, writing in June 1544, "Tolosani's
Condemnations of Copernicus' *Revolutions,*" in *Copernicus and the
Scientific Revolution,* ed. Edward Rosen (Malabar, Fla.: Krieger,
1984), 189.

8. Georg Joachim Rheticus, *First Account* (1540), in *Three Copernican Treatises,* trans. Edward Rosen, 3d ed. (New York: Octagon
Books, 1971), 139; Copernicus, *Revolutions,* 1:10.

9. Galileo Galilei, *Sidereus Nuncius* (Venice, 1610), folios 15r and
16r, quoted from *The Book of the Cosmos,* ed. Danielson, 149–50; my

translation adapted from *The Sidereal Messenger of Galileo Galilei*, trans. E. S. Carlos (London: Dawsons, 1880). See also *Sidereus Nuncius: or The Sidereal Messenger*, trans. Albert van Helden (Chicago: University of Chicago Press, 1989), 55–57.

10. *Kepler's Conversation with Galileo's Sidereal Messenger* (1610), trans. Edward Rosen (New York: Johnson Reprint Corporation, 1965), 45–46 (italics added).

11. John Wilkins, *The Mathematical and Philosophical Works* (London: Frank Cass, 1970), 190–91.

12. John Donne, *An Anatomy of the World* (London, 1611); Blaise Pascal, *Pensées* (ca. 1650), from *Thoughts*, trans. W. F. Trotter (New York: Collier, 1910), 78; Cotton Mather, *The Christian Philosopher: A Collection of the Best Discoveries in Nature, with Religious Improvements* (London, 1721), 19.

13. Cyrano de Bergerac, *The Government of the World in the Moon* (London, 1659), sig. B8v; Bernard Le Bouvier de Fontenelle, *Entretiens sur la Pluralité des Mondes* (Paris, 1686), trans. W. D. Knight as *A Discourse of the Plurality of Worlds* (Dublin, 1687), 11–13; Johann Wolfgang Goethe, *Materialien zur Geschichte der Farbenlehre*, in *Goethes Werke*, Hamburger Ausgabe (Hamburg: Christian Wegner Verlag, 1960), 14:81.

MYTH 7: THAT GIORDANO BRUNO WAS THE FIRST MARTYR OF MODERN SCIENCE

Epigraphs: Michael White, *The Pope and the Heretic: The True Story of Giordano Bruno, the Man Who Dared to Defy the Roman Inquisition* (New York: HarperCollins, 2002), 3. Giordano Bruno, *The Ash Wednesday Supper: La Cena de le ceneri*, ed. and trans. Edward A. Gosselin and Lawrence S. Lerner (Hamden, Conn.: Shoestring Press, 1977), 11–12.

1. Andrew Dickson White, *The Warfare of Science* (New York: D. Appleton, 1876). See also John William Draper, *History of the Conflict between Religion and Science* (New York: D. Appleton, 1874), 178–80.

2. On the role of White and Draper in discussions of the hostility of Christian churches and theology toward science, see David C. Lindberg, "Science and the Early Church," in *God and Nature: Historical Essays on the Encounter between Christianity and Science*, ed. David C. Lindberg and Ronald L. Numbers (Berkeley: University of California Press, 1986), 19–48, esp. 19–22; see also the editors' introduction, esp. 1–4.

3. Hugh Kearney, *Science and Change 1500–1700* (New York: McGraw-Hill, 1971), 106.

4. See, e.g., W. P. D. Wightman, *Science in a Renaissance Society* (London: Hutchinson, 1972), 127–28.

5. This is the view of Angelo Mercati, *Il Sommario del Processo di Giordano Bruno, con Appendice di Documenti sull' Eresia e l' Inquisizione a Modena nel Secolo XVI* (Vatican City: Biblioteca Apostolica Vaticana, 1942), summarized by Frances Yates, *Giordano Bruno and the Hermetic Tradition* (Chicago: University of Chicago Press, 1964), 354. This view is still held by Richard Olson, *Science and Religion 1450–1900: From Copernicus to Darwin* (Westport, Conn.: Greenwood Press, 2004), 58.

6. Ramon G. Mendoza, *The Acentric Labyrinth: Giordano Bruno's Prelude to Contemporary Cosmology* (Shaftesbury, Dorset: Element Books, 1995), 52–53, identifies Mercati as a member of the Papal Curia who was motivated to declassify the summary document by his wish "to exonerate the Pope and the Roman Inquisition from blame in the trial and execution of Giordano Bruno."

7. William F. Bynum, "The Great Chain of Being," in *The History of Science and Religion in the Western Tradition: An Encyclopedia,* ed. Gary B. Ferngren (New York: Garland, 2000), 444–446, 445 (emphasis added).

8. White, *Pope and Heretic,* dust jacket (emphasis added).

9. Hilary Gatti, "The Natural Philosophy of Giordano Bruno," *Midwest Studies in Philosophy* 26 (2002): 111–23, and Karen Silvia de León-Jones, *Giordano Bruno and the Kabbalah: Prophets, Magicians, and Rabbis* (Lincoln: University of Nebraska Press, 1997), 2–5, give excellent surveys of Bruno's fortunes in the European philosophical tradition.

10. Bertolt Brecht's play *Galileo* clearly identifies the Italian astronomer as a seeker of truth muzzled by the censors of a totalitarian regime and Bruno as his harbinger. See Bertolt Brecht, *Galileo,* ed. Eric Bentley, trans. Charles Laughton (New York: Grove Press, 1966), 62–63.

11. Volker R. Remmert, " 'Docet parva pictura, quod multae scripturae non dicunt': Frontispieces, Their Functions, and Their Audiences in Seventeenth-Century Mathematical Sciences," in *Transmitting Knowledge: Words, Images, and Instruments in Early Modern Europe,* ed. Sachiko Kusukawa and Ian Maclean (Oxford: Oxford University Press, 2006), 239–70, esp. 250–56. Remmert's insights

build on material presented by William B. Ashworth, Jr., in various public talks and published in "Divine Reflections and Profane Refractions: Images of a Scientific Impasse in 17th-Century Italy," in *Gianlorenzo Bernini: New Aspects of His Art and Thought*, ed. Irving Lavin (University Park: Pennsylvania State University Press, 1985), 179–207; "Allegorical Astronomy: Baroque Scientists Encoded Their Most Dangerous Opinions in Art," *The Sciences* 25 (1985): 34–37; and a forthcoming book, *Emblematic Imagery of the Scientific Revolution*.

12. On Kepler's comment, see Hilary Gatti, *Giordano Bruno and Renaissance Science* (Ithaca, N.Y.: Cornell University Press, 1999), 56.

13. Ludovico Geymonat, *Galileo Galilei: A Biography and Inquiry into His Philosophy of Science*, trans. Stillman Drake (New York: McGraw Hill, 1965), 60. See also Giorgio de Santillana, *The Crime of Galileo* (Chicago: University of Chicago Press, 1955), 26; and Howard Margolis, *It Started with Copernicus: How Turning the World Inside Out Led to the Scientific Revolution* (New York: McGraw Hill, 2002), 149. The notion of Bruno's "ghost" is manifest in the title of the popular science article by Lawrence S. Lerner and Edward A. Gosselin, "Galileo and the Specter of Giordano Bruno," *Scientific American* 255, no. 5 (1986): 126–33.

14. The summary document of Bruno's trial before the Roman Inquisition (*Sommario del processo*, Rome, 1 March 1598) is printed as document 51 in Luigi Firpo, *Il Processo di Giordano Bruno* (Rome: Salerno Editrice, 1993), 247–304.

15. Giordano Bruno, *The Cabala of Pegasus*, trans. Sidney L. Sondergard and Madison U. Sowell (New Haven, Conn.: Yale University Press, 2002), 6.

16. Ibid., xviii: Bruno identified himself as *sacrae theologiae professor* when presenting himself to the Academy at Geneva in 1579.

17. This point of view, critical of the failure of positivist historiography to grasp the unity of Bruno's intellectual program, is clearly expressed by Gatti, "Natural Philosophy of Giordano Bruno," 118, and by De León-Jones, *Giordano Bruno and the Kabbalah*, 5. See also Yates, *Giordano Bruno*, 366; and Dorothea Waley Singer, *Giordano Bruno: His Life and Thought, with Annotated Translation of His Work* On the Infinite Universe and Worlds (New York: Schuman, 1950), 165.

MYTH 8: THAT GALILEO WAS IMPRISONED AND TORTURED FOR ADVOCATING COPERNICANISM

Epigraphs: Voltaire, "Descartes and Newton," in *Essays on Literature, Philosophy, Art, History,* in *The Works of Voltaire,* trans. William F. Fleming, ed. Tobis Smollett et al., 42 vols. (Paris: Du Mont, 1901), 37:167. Giuseppe Baretti, *The Italian Library* (London, 1757), 52. Italo Mereu, *Storia dell'intolleranza in Europa* (Milan: Mondadori, 1979), 385. A version of the prison myth can also be found in the context of recent popular culture. It occurs in the PBS two-hour program titled *Galileo's Battle for the Heavens,* first aired on 29 October 2002. Near the end of the program, there is a scene in which Galileo arrives at his house in Arcetri after the condemnation, and the door to the house is shown being closed and locked with a key from the outside. This suggests that he was not free to go in and out of his house, whereas in fact he was: he was free to walk in the villa's gardens and to travel the few blocks to the nearby convent where his daughter was a nun.

1. For other statements of the prison thesis, see Luc Holste to Nicolas de Peiresc (7 March [i.e., May] 1633), in Galileo Galilei, *Opere,* ed. A. Favaro et al., 20 vols. (Florence: Barbèra, 1890–1909), 15:62; John Milton, *Areopagitica* (London, 1644), 24; Domenico Bernini, *Historia di tutte l'heresie,* 4 vols. (Rome, 1709), 4:615; Louis Moreri, *Le grand dictionnaire historique,* 5 vols. (Paris, 1718), 1:196; Jean B. Delambre, *Histoire de l'astronomie moderne,* 2 vols. (Paris, 1821), 1:671; John William Draper, *History of the Conflict between Religion and Science* (New York: D. Appleton, 1874), 171–72; E. H. Haeckel, *Gesammelte populäre Vorträge aus dem Gebiete der Entwicklungslehre* (Bonn, 1878–1879), 33; Andrew D. White, *A History of the Warfare of Science with Theology in Christendom,* 2 vols. (New York, 1896), 2:142; and Bertrand Russell, *Religion and Science* (Oxford: Oxford University Press, 1935), 40. On the torture thesis, see also Paolo Frisi, *Elogio del Galileo* (Milan, 1775), in *Elogi: Galilei, Newton, D'Alembert,* ed. Paolo Casini (Rome: Theoria, 1985), 71; Giovanni B. C. Nelli, *Vita e commercio letterario di Galileo Galilei,* 2 vols. (Lausanne [i.e., Florence], 1793), 2:542–54; Guglielmo Libri, *Essai sur la vie et les travaux de Galilée* (Paris, 1841), 34–37; Silvestro Gherardi, *Il processo di Galileo riveduto sopra documenti di nuova fonte* (Florence, 1870), 52–54; Emil Wohlwill, *Ist Galilei gefoltert worden?* (Leipzig, 1877); J. A. Scartazzini, "Il processo di Galileo Galilei e la moderna critica tedesca,"

Rivista europea 4 (1877): 821–61, 5 (1878): 1–15, 5 (1878): 221–49, 6 (1878): 401–23, and 18 (1878): 417–53; and Enrico Genovesi, *Processi contro Galileo* (Milan: Ceschina, 1966), 232–82.

2. The simplified account of Galileo's trial given in the preceding paragraphs is distilled from the following standard works: Giorgio de Santillana, *The Crime of Galileo* (Chicago: University of Chicago Press, 1955); Stillman Drake, *Galileo at Work* (Chicago: University of Chicago Press, 1978); Maurice A. Finocchiaro, *Galileo and the Art of Reasoning* (Boston: Dordrecht, 1980); Maurice A. Finocchiaro, trans. and ed., *The Galileo Affair: A Documentary History* (Berkeley: University of California Press, 1989); Mario Biagioli, *Galileo, Courtier* (Chicago: University of Chicago Press, 1993); Rivka Feldhay, *Galileo and the Church* (Cambridge: Cambridge University Press, 1995); Massimo Bucciantini, *Contro Galileo* (Florence: Olschki, 1995); Francesco Beretta, *Galilée devant le Tribunal de l'Inquisition* (doctoral diss., Faculty of Theology, University of Fribourg, Switzerland, 1998); Annibale Fantoli, *Galileo: For Copernicanism and for the Church*, trans. George V. Coyne, 3d ed. (Vatican City: Vatican Observatory Publications, 2003); William R. Shea and Mariano Artigas, *Galileo in Rome* (Oxford: Oxford University Press, 2003); Michele Camerota, *Galileo Galilei e la cultura scientifica nell'età della Controriforma* (Rome: Salerno Editrice, 2004); Ernan McMullin, ed., *The Church and Galileo* (Notre Dame, Ind.: University of Notre Dame Press, 2005); Mario Biagioli, *Galileo's Instruments of Credit* (Chicago: University of Chicago Press, 2006); Richard J. Blackwell, *Behind the Scenes at Galileo's Trial* (Notre Dame, Ind.: University of Notre Dame Press, 2006); Antonio Beltrán Marí, *Talento y poder: Historia de las relaciones entre Galileo y la Iglesia católica* (Pamplona: Laetoli, 2006).

3. Quoted from Finocchiaro, *Galileo Affair*, 290, 291; see also Galilei, *Opere*, 19:405, 406.

4. Galilei, *Opere*, 15:169, 19:411–15; Maurice A. Finocchiaro, *Retrying Galileo, 1613–1992* (Berkeley: University of California Press, 2005), 26–42.

5. Eliseo Masini, *Sacro arsenale overo prattica dell'officio della Santa Inquisitione* (Genoa, 1621), 121–51; Desiderio Scaglia, *Prattica per proceder nelle cause del Santo Uffizio*, unpublished manuscript, ca. 1615–1639 (?), now available in Alfonso Mirto, "Un inedito del Seicento sull'Inquisizione," *Nouvelles de la république des lettres*, 1986, no. 1, 99–138, at 133; Nicola Eymerich and Francisco Peña, *Directorium inquisitorum* (Rome, 1578), now in *Le manuel des inquisiteurs*, ed. and trans. Louis Sala-Molins (Paris: Mouton, 1973), 158–64, 207–12; Beretta, *Galilée devant le Tribunal de l'Inquisition*, 214–21.

6. Some crucial letters were published in Florence in 1774, as reported in Nelli, *Vita e commercio letterario di Galileo Galilei*, 2:537–38. A larger collection was published in Angelo Fabroni, ed., *Lettere inedite di uomini illustri*, vol. 2 (Florence, 1775). A good and accurate summary of this evidence was given by Girolamo Tiraboschi, "Sulla condanna del Galileo e del sistema copernicano" (lecture read at the Accademia de' Dissonanti, Modena, 7 March 1793), in *Storia della letteratura italiana* (Rome, 1782–1797), 10:373–83, at 382, translated in Finocchiaro, *Retrying Galileo*, 171. The 1633 correspondence can now be found in Galilei, *Opere*, vol. 15; the trial depositions in 19:336–62. Translations of the most important correspondence and of all four depositions are given in Finocchiaro, *Galileo Affair*, 241–55, 256–87, respectively. For the distinction between the Tuscan embassy (Palazzo Firenze) and Villa Medici, see Shea and Artigas, *Galileo in Rome*, 30, 74, 106–7, 134–35, 179–80, 195.

7. The trial proceedings were published by Henri de L'Epinois, "Galilée: Son Procès, Sa Condamnation d'après des Documents Inédits," *Revue des questions historiques*, year 2, vol. 3 (1867): 68–171; Domenico Berti, *Il processo originale di Galileo Galilei pubblicato per la prima volta* (Rome, 1876); Karl von Gebler, *Die Acten des Galilei'schen Processes, nach der Vaticanischen Handschrift* (Stuttgart, 1877); Henri de L'Epinois, *Les pièces du procès de Galilée précédées d'un avant-propos* (Paris, 1877); Domenico Berti, *Il processo originale di Galileo Galilei: Nuova edizione accresciuta, corretta e preceduta da un'avvertenza* (Rome, 1878). Besides the proponents of the torture thesis mentioned in note 4, essential works in the assimilation process included Marino Marini, *Galileo e l'Inquisizione* (Rome, 1850), 54–68; Th. Henri Martin, *Galilée, les droits de la science et la méthode des sciences physiques* (Paris, 1868), 123–31; Sante Pieralisi, *Urbano VIII e Galileo Galilei* (Rome, 1875), 227–46; Berti, *Processo originale di Galileo* (1876), cv–cxvii; Henri de L'Epinois, *La question de Galilée* (Paris, 1878), 197–216; Karl von Gebler, *Galileo Galilei and the Roman Curia*, trans. Mrs. George Sturge (London, 1879), 252–63; Léon Garzend, "Si l'Inquisition avait, en principe, décidé de torturer Galilée?" *Revue pratique d'apologétique* 12 (1911): 22–38, 265–78; Léon Garzend, "Si Galilée pouvait, juridiquement, etre torturé," *Revue des questions historiques* 90 (1911): 353–89, and 91 (1912): 36–67; and Orio Giacchi, "Considerazioni giuridiche sui due processi contro Galileo," in *Nel terzo centenario della morte di Galileo Galilei*, ed. Università Cattolica del Sacro Cuore (Milan: Società Editrice "Vita e Pensiero," 1942), 383–406.

8. The first is quoted from Finocchiaro, *Retrying Galileo*, 246; see also Galilei, *Opere*, 19:282–83; and Epinois, "Galilée: Son procès," 129, n. 4. The other is from Finocchiaro, *Galileo Affair*, 287; see also Galilei, *Opere*, 19:362.

9. See Scaglia, *Prattica per proceder nelle cause del Santo Uffizio*, 133; Genovesi, *Processi contro Galileo*, 79–81; Mereu, *Storia dell'intolleranza in Europa*, 226–27; Beretta, *Galilée devant le Tribunal*, 216; and Beltrán Marí, *Talento y poder*, 797.

10. See Masini, *Sacro arsenale*, 120–51; Berti, *Processo originale di Galileo*, cv–cxvii; Gebler, *Galileo Galilei and the Roman Curia*, 256–57; and Beretta, *Galilée devant le Tribunal*, 214–21.

11. See especially Garzend, "Si Galilée pouvait, juridiquement, etre torturé," citing an impressive array of treatises on canon law, civil law, theology, and inquisitorial practice from Galileo's time. For Galileo's clerical tonsure, see Galilei, *Opere*, 19:579–80.

12. On abuses of the system that occurred in 1604 in the Paduan Inquisition's investigations of Galileo and Cesare Cremonini, see Beltrán Marí, *Talento y poder*, 25–45.

13. On degrees of torture, see Masini, *Sacro arsenale*, 120–51; Philippus van Limborch, *Historia Inquisitionis* (Amsterdam, 1692), 322; Scartazzini, "Il processo di Galileo Galilei," 6:403–4; Gebler, *Galileo Galilei and the Roman Curia*, 256, n. 2; Genovesi, *Processi contro Galileo*, 252–55; Eymerich and Peña, *Le manuel des inquisiteurs*, 209; and the original sources to which most of these authors refer: Paolo Grillandi, *Tractatus de questionibus et tortura* (1536), question 4, number 11; and Julius Clarus, *Practica criminalis* (Venice, 1640), question 64. For support of the *territio-realis* thesis, see Wohlwill, *Ist Galilei gefoltert worden?* 25–28. For criticism, see Gebler, *Galileo Galilei and the Roman Curia*, 254–56; and Finocchiaro, *Retrying Galileo*, 252.

14. Proponents of the moral-torture thesis include Jean Biot, *Mélanges scientifiques et littéraires*, 3 vols. (Paris, 1858), 3:42–43; Philarète Chasles, *Galileo Galilei: Sa vie, son procès et ses contemporaines* (Paris, 1862); Joseph L. Trouessart, *Galilée: Sa mission scientifique, sa vie et son procès* (Poitiers, 1865), 110. For criticism of the moral-torture thesis, see Pieralisi, *Urbano VIII e Galileo Galilei*, 242–46. See also Finocchiaro, *Retrying Galileo*, 234, 236.

MYTH 9: THAT CHRISTIANITY GAVE BIRTH TO MODERN SCIENCE

Epigraphs: Alfred North Whitehead, *Science and the Modern World* (New York: Macmillan, 1925), 19. Stanley L. Jaki, *The Road of Science and the Ways to God* (Chicago: University of Chicago Press, 1978), 243. Rodney Stark, *For the Glory of God: How Monotheism Led to Reformations, Science, Witch-hunts and the End of Slavery* (Princeton, N.J.: Princeton University Press, 2003), 3, 123 (italics in original). Other versions of this view include Reijer Hooykaas, *Religion and the Rise of Modern Science* (Grand Rapids, Mich.: Eerdmans, 1972), and Jaki, *The Road of Science.*

1. Stark, *Glory of God,* 233.

2. For an excellent short survey of the complicated and varied relationships between Christianity and natural philosophy before the Enlightenment, see David C. Lindberg and Peter Harrison, "Science and the Christian Church: From the Advent of Christianity to 1700," in John Hedley Brooke and Ronald L. Numbers, eds., *Science and Religion around the World: Historical Perspectives* (New York: Oxford University Press, forthcoming).

3. Robert King Merton, *Science, Technology and Society in Seventeenth Century England,* originally published in the journal *Osiris,* 4, pt. 2 (Bruges: St. Catherine Press, 1938): 360–632. Among the antecedent studies Merton cites is Alphonse de Candolle, *Histoire des sciences et des savants depuis deux siècles: suivie d'autres études sur des sujets scientifiques en particulier sur la sélection dans l'espèce humaine* (Genève: H. Georg, 1873).

4. For a discussion of Merton's theory, see the essays in I. Bernard Cohen, K. E. Duffin, and Stuart Strickland, *Puritanism and the Rise of Modern Science: The Merton Thesis* (New Brunswick, N.J.: Rutgers University Press, 1990).

5. Earlier, canonical statements of this view can be found in M. B. Foster, "The Christian Doctrine of Creation and the Rise of Modern Natural Science," *Mind* 18 (1934): 446–68; and Francis Oakley, "Christian Theology and the Newtonian Science: The Rise of the Concept of Laws of Nature," *Church History* 30 (1961): 433–57. For marvelous, more recent discussions, see John Henry, "Metaphysics and the Origins of Modern Science: Descartes and the Importance of Laws of Nature," *Early Science and Medicine* 9 (2004): 73–114; and Peter Harrison, "The Development of the Concept of Laws of Nature," in *Creation: Law and Probability,* ed. Fraser Watts (Aldershot: Ashgate, 2007).

6. On this theme, see Peter Harrison, *The Fall of Man and the Foundations of Science* (Cambridge: Cambridge University Press, 2008).

7. Peter Harrison, *The Bible, Protestantism, and the Rise of Natural Science* (Cambridge: Cambridge University Press, 1998).

8. Descartes to William Boswell, 1646, *Oeuvres de Descartes,* ed. Charles Adam and P. Tannery, 11 vols. (Paris: Cerf, 1897–1913), 4:698.

9. J. E. McGuire and P. M. Rattansi, "Newton and the 'Pipes of Pan,' " *Notes and Records of the Royal Society of London* 21 (1966): 108–43.

10. John Heilbron has recently argued that "the Roman Catholic Church gave more financial and social support to the study of astronomy for over six centuries, from the recovery of ancient learning during the late Middle Ages into the Enlightenment, than any other, and, probably, all other, institutions." See Heilbron, *The Sun in the Church: Cathedrals as Solar Observatories* (Cambridge, Mass.: Harvard University Press, 1999), 3. For Jesuit contributions to seventeenth-century science, see Mordechai Feingold, ed., *The New Science and Jesuit Science: Seventeenth-Century Perspectives* (Dordrecht: Kluwer, 2003).

11. A remarkable, brief description of this history can be found in David C. Lindberg, "The Medieval Church Encounters the Classical Tradition," in *When Science and Christianity Meet,* ed. David C. Lindberg and Ronald L. Numbers (Chicago: University of Chicago Press, 2003), 7–32.

12. For a richer description, see Myth 4 in this book, "That Medieval Islamic Culture Was Inhospitable to Science."

13. David C. Lindberg, ed., *Science in the Middle Ages* (Chicago: University of Chicago Press, 1978), 13. See also Myth 4 in this book.

14. See Edward Grant, *A History of Natural Philosophy: From the Ancient World to the Nineteenth Century* (Cambridge: Cambridge University Press, 2007), 130–78.

15. George Saliba, *Islamic Science and the Making of the European Renaissance* (Cambridge, Mass.: MIT Press, 2007), 221. For another effort to trace links between late-medieval Islamic and early-modern Christian astronomy, see F. Jamil Ragep, "Ali Qushi and Regiomontanus: Eccentric Transformations and Copernican Revolutions," *Journal for the History of Astronomy* 36 (2005): 359–71. Also see the discussion of the role of Ibn al-Haytham's optics on Kepler and Galileo in David J. Hess, *Science and Technology in a Multicultural World: The Cultural Politics of Facts and Artifacts* (New York: Columbia University Press, 1995), 66.

16. Clifford Geertz, "Thick Description: Toward an Interpretive Theory of Culture," in *The Interpretation of Cultures: Selected Essays,* ed. Clifford Geertz (New York: Basic Books, 1973), 28–29.

17. Arun Bala, *The Dialogue of Civilizations in the Birth of Modern Science* (New York: Palgrave Macmillan, 2006).

18. Harold J. Cook, *Matters of Exchange: Commerce, Medicine and Science in the Dutch Golden Age* (New Haven, Conn.: Yale University Press, 2007).

19. See, for example, David Landes, *Revolutions in Time: Clocks and the Making of the Modern World* (Cambridge, Mass.: Belknap Press, 1983); Elisabeth Eisenstein, *The Printing Press as an Agent of Change* (Cambridge: Cambridge University Press, 1979); and Tal Golan, *Laws of Nature and Laws of Man* (Cambridge, Mass.: Harvard University Press, 2004).

20. Thomas Sprat, *History of the Royal Society*, ed. J. Cope and H. W. Jones (London: Routledge and Kegan Paul, 1959), 63. Sprat's account was first published in 1667.

21. Jonathan I. Israel, *Enlightenment Contested* (New York: Oxford University Press, 2006), 201-22, 356-71. Also see Bernard Lightman, "Science and Unbelief," in Brooke and Numbers, eds., *Science and Religion around the World*.

22. Robert K. Merton, "Science and the Social Order," *Philosophy of Science 5* (1938): 321-37. This "ethos" itself has its own cultural history, described preliminarily and brilliantly in Stephen Gaukroger, *The Emergence of a Scientific Culture* (Oxford: Oxford University Press, 2006).

23. Rodney Stark, "False Conflict," *The American Enterprise* October/November (2003): 27.

24. Ibid., 14.

25. Lynn White, Jr., "The Historical Roots of Our Ecological Crisis," *Science* 155 (1967): 1203-7.

MYTH 10: THAT THE SCIENTIFIC REVOLUTION LIBERATED SCIENCE FROM RELIGION

Epigraphs: Richard S. Westfall, "The Scientific Revolution Reasserted," in *Rethinking the Scientific Revolution*, ed. Margaret J. Osler (Cambridge: Cambridge University Press, 2000), 41-55, at 49-50. Jonathan I. Israel, *Radical Enlightenment: Philosophy and the Making of Modernity, 1650-1750* (Oxford: Oxford University Press, 2001), 14. I am grateful to John H. Brooke, Peter Harrison, Ronald L. Numbers, and Michael Shank for helpful comments on an earlier version of this chapter.

1. John Locke, *An Essay Concerning Human Understanding*, ed. Peter H. Nidditch (Oxford: Clarendon Press, 1975), bk. 4, chap. 12, §10, p. 645.

2. John Hedley Brooke, *Science and Religion: Some Historical Perspectives* (Cambridge: Cambridge University Press, 1991), chap. 2.

3. Galileo Galilei, "Letter to the Grand Duchess," in Stillman Drake, *Discoveries and Opinions of Galileo* (Garden City, N.Y.: Doubleday, 1957), 182.

4. Robert Boyle, *Excellency of Theology: Or,The Preeminence of the Study of Divinity, above that of Natural Philosophy,* in *The Works of Robert Boyle,* ed. Michael Hunter and Edward B. Davis, 11 vols. (London: Pickering and Chatto, 2000), 8:27.

5. Isaac Newton, *The Principia: Mathematical Principles of Natural Philosophy,* trans. I. Bernard Cohen and Anne Whitman (Berkeley: University of California Press, 1999), 942–43.

MYTH 11: THAT CATHOLICS DID NOT CONTRIBUTE TO THE SCIENTIFIC REVOLUTION

Epigraphs: John William Draper, *A History of the Conflict between Religion and Science* (New York: D. Appleton, 1874), 363–64. Charles C. Gillispie, *The Edge of Objectivity: An Essay in the History of Scientific Ideas* (Princeton, N.J.: Princeton University Press, 1960), 114.

1. Donald Fleming, *John William Draper and the Religion of Science* (Philadelphia: University of Pennsylvania Press, 1950), 31. Contemporaneous reviews of Draper's book were very mixed; the periodical *Catholic World* 21 (1875): 178–200 saw it simply as the "latest addition to anti-Catholic literature" and called it "a farrago of falsehoods, with an occasional ray of truth, all held together by the slender thread of a spurious philosophy."

2. Philip Jenkins, *The New Anti-Catholicism: The Last Acceptable Prejudice* (New York: Oxford University Press, 2005), 23.

3. An early example is in Robert Boyle's 1648 autobiographical memoir "An Account of Philaretus during His Minority" in *Robert Boyle by Himself and His Friends,* ed. Michael Hunter (London: Pickering and Chatto, 1994), 1–22, on 19–21.

4. In the case of astronomy, for example, see J. L. Heilbron, *The Sun in the Church: Cathedrals as Solar Observatories* (Cambridge, Mass.: Harvard University Press, 1999), 3: "The Roman Catholic Church gave more financial and social support to the study of astronomy for over six centuries, from the recovery of ancient learning during the late Middle Ages into the Enlightenment, than any other, and, probably, all other, institutions." See also William B. Ashworth, Jr., "Catholicism and Early Modern Science," in *God and Nature: A History of the Encounter*

between Christianity and Science, ed. David C. Lindberg and Ronald L. Numbers (Berkeley: University of California Press, 1986), 136–66.

5. Walter Pagel, *Joan Baptista Van Helmont: Reformer of Science and Medicine* (Cambridge: Cambridge University Press, 1982); Pagel, *Religious and Philosophical Aspects of Van Helmont's Science and Medicine, Bulletin of the History of Medicine,* supp. 2, 1944.

6. Alan Cutler, *The Seashell on the Mountaintop: A Story of Science, Sainthood, and the Humble Genius Who Discovered a New History of the Earth* (New York: Dutton, 2003).

7. Peter Dear, *Mersenne and the Learning of the Schools* (Ithaca, N.Y.: Cornell University Press, 1988).

8. W. E. Knowles Middleton, *The Experimenters: A Study of the Accademia del Cimento* (Baltimore: Johns Hopkins University Press, 1971); David Freedberg, *The Eye of the Lynx: Galileo, His Friends, and the Beginnings of Modern Natural History* (Chicago: University of Chicago Press, 2002); David J. Sturdy, *Science and Social Status: The Members of the Académie des Sciences, 1666–1750* (Rochester, N.Y.: Boydell Press, 1995).

9. Steven J. Harris, "Jesuit Scientific Activity in the Overseas Missions, 1540–1773," *Isis* 96 (2005): 71–79.

10. *Jesuit Science and the Republic of Letters,* ed. Mordechai Feingold (Cambridge, Mass.: MIT Press, 2003); *The Jesuits: Cultures, Sciences, and the Arts, 1540–1773,* ed. John O'Malley, Gauvin Alexander Bailey, Steven J. Harris, and T. Frank Kennedy (Toronto: University of Toronto Press, 1999) and *The Jesuits II: Cultures, Sciences, and the Arts, 1540–1773* (Toronto: University of Toronto Press, 2006); Agustín Udías, *Searching the Heavens and the Earth: The History of Jesuit Observatories* (Dordrecht: Kluwer, 2003); Marcus Hellyer, *Catholic Physics: Jesuit Natural Philosophy in Early Modern Germany* (Notre Dame, Ind.: University of Notre Dame Press, 2005); William A. Wallace, *Galileo and His Sources: The Heritage of the Collegio Romano in Galileo's Science* (Princeton, N.J.: Princeton University Press, 1984).

11. Mark A. Waddell, "The World, As It Might Be: Iconography and Probablism in the *Mundus Subterraneus* of Athanasius Kircher," *Centaurus* 48 (2006): 3–22; Paula Findlen, *Athanasius Kircher: The Last Man Who Knew Everything* (New York: Routledge, 2004).

12. See, for example, Edward Grant, *The Foundations of Modern Science in the Middle Ages* (Cambridge: Cambridge University Press, 1996) and David A. Lindberg, *The Beginnings of Western Science,* 2d ed. (Chicago: University of Chicago Press, 2007), 357–67.

MYTH 12: THAT RENÉ DESCARTES ORIGINATED THE MIND-BODY DISTINCTION

Epigraph: Antonio Damasio, *Descartes' Error: Emotion, Reason and the Human Brain* (New York: Quill, 1994), 249–50.

1. Stephen Gaukroger, *Descartes: An Intellectual Biography* (Oxford: Oxford University Press, 1995), 2.

2. Gilbert Ryle, *The Concept of Mind* (London: Hutchinson, 1949). See also Richard A. Watson, "Shadow History in Philosophy," *Journal of the History of Philosophy* 31 (1993): 95–109 (esp. 102–5).

3. Daniel Dennett, *Consciousness Explained* (London: Penguin, 1993), 39; review of Antonio Damasio, *Descartes' Error: Emotion, Reason, and the Human Brain, Times Literary Supplement,* 25 August 1995, 3–4, at 3.

4. Dennett, *Consciousness Explained,* 37.

5. Damasio, *Descartes' Error,* 249–50.

6. Descartes, *Meditations* §78, in *The Philosophical Writings of Descartes,* trans. John Cottingham, Robert Stoothoff, and Dugald Murdoch, 3 vols. (Cambridge: Cambridge University Press, 1984–1991), 2:54; cf. *Principles of Philosophy* §25, *Philosophical Writings,* 1:210.

7. Descartes, *Meditations* §81, in *Philosophical Writings,* 2:56; cf. *Discourse on the Method* §59, in *Philosophical Writings,* 1:141; Descartes, *Objections and Replies* §227, in *Philosophical Writings,* 2:160.

8. Aristotle, *De anima* 413a8; Plotinus, *Enneads* IV.iii.21; Aquinas, *Questions on the Soul,* Q11; *On the Power of God,* bk. 2, Q5, A1; *On Spiritual Creatures,* A2; *Treatise on Separate Substances,* chap. 1, sec. 7.

9. D. J. O'Meara, *Plotinus: An Introduction to the Enneads* (Oxford: Oxford University Press, 1993), 19–20.

10. Descartes, *Passions of the Soul,* §342, in *Philosophical Writings,* 1:335. See also Deborah Brown, *Descartes and the Passionate Mind* (Cambridge: Cambridge University Press, 2006).

11. See Gary Hatfield, "The *Passions of the Soul* and Descartes' Machine Psychology," *Studies in History and Philosophy of Science* 38 (2007): 1–35.

12. Gordon Baker and Katherine J. Morris, *Descartes' Dualism* (London: Routledge, 1996). See also Stephen Nadler's illuminating review in *Philosophical Books* 37 (1998): 157–69.

13. Descartes to Elizabeth, 21 May 1643, in *Descartes: Philosophical Letters,* ed. Anthony Kenny (Oxford: Clarendon, 1970), 138.

14. John Cottingham, "Cartesian Trialism," *Mind* 94 (1985): 218–30. Cf. Peter Remnant, "Descartes: Body and Soul," *Canadian Journal of Philosophy* 9 (1979): 377–86; Paul Hoffman, "The Unity of Descartes's Man," *Philosophical Review* 95 (1986): 339–70.

15. Steven Nadler, "Descartes and Occasional Causation," *British Journal for the History of Philosophy* 2 (1994): 35–54, and the essays in Steven Nadler, ed., *Causation in Early Modern Philosophy* (University Park: Penn State Press, 1993).

16. David Yandell, "What Descartes Really Told Elizabeth: Mind-Body Union as Primitive Notion," *British Journal for the History of Philosophy* 5 (1997): 249–73; Baker and Morris, *Descartes' Dualism*, 154.

17. Ryle, *Concept of Mind*, 23.

MYTH 13: THAT ISAAC NEWTON'S MECHANISTIC COSMOLOGY ELIMINATED THE NEED FOR GOD

I am grateful to Lawrence Principe, Margaret Osler, and Stephen Snobelen for helpful comments.

Epigraphs: http://en.wikipedia.org/wiki/Clockwork_universe_theory, (accessed 11 June 2007). Sylvan S. Schweber, "John Herschel and Charles Darwin: A Study in Parallel Lives," *Journal of the History of Biology* 22 (1989): 1–71, on 1. Thomas H. Greer, *A Brief History of the Western World*, 4th ed. (New York: Harcourt Brace Jovanovich, 1982), 364.

1. For a longer version of these ideas, see Edward B. Davis, "Newton's Rejection of the 'Newtonian World View': The Role of Divine Will in Newton's Natural Philosophy," in *Facets of Faith and Science, Volume 3: The Role of Beliefs in the Natural Sciences*, ed. Jitse M. van der Meer (Lanham, Md.: University Press of America, 1996), 75–96; reprinted with additions from *Science & Christian Belief* 3 (1991): 103–117, and *Fides et Historia* 22 (Summer 1990): 6–20.

2. Westfall stated this most forcefully in *Science and Religion in Seventeenth-Century England* (New Haven, Conn.: Yale University Press, 1958), chap. 8. His views were not much different at the time of his death; see Westfall, "Isaac Newton," in *The History of Science and Religion in the Western Tradition: An Encyclopedia*, ed. Gary B. Ferngren, Edward J. Larson, and Darrel W. Amundsen (New York: Garland Publishing, 2000), 95–99.

3. Betty Jo Teeter Dobbs, *The Janus Faces of Genius: The Role of Alchemy in Newton's Thought* (Cambridge: Cambridge University Press, 1991).

4. Yahuda MS 15, fols. 96r/v, Jewish National and University Library, Jerusalem, http://www.newtonproject.sussex.ac.uk/texts/viewtext.php?id =THEM00222&mode=normalized (accessed 7 March 2008). Newton's reference here to "1 John" means the first chapter of John's gospel, not John's first epistle.

5. Stephen D. Snobelen, " 'God of Gods, and Lord of Lords': The Theology of Isaac Newton's General Scholium to the *Principia*," *Osiris* 16 (2001): 169–208.

6. Isaac Newton, *The Principia: Mathematical Principles of Natural Philosophy*, a new translation by I. Bernard Cohen and Anne Whitman, assisted by Julia Budenz (Berkeley: University of California Press, 1999), 940–41.

7. *The Leibniz-Clarke Correspondence*, ed. H. G. Alexander (Manchester: Manchester University Press, 1956), 11–12 and 14.

MYTH 14: THAT THE CHURCH DENOUNCED ANESTHESIA IN CHILDBIRTH ON BIBLICAL GROUNDS

Epigraphs: Andrew Dickson White, *A History of the Warfare of Science with Theology in Christendom*, 2 vols. (New York: D. Appleton, 1896), 2:63. Deborah Blum, "A Pox on Stem Cell Research," Op-Ed, *New York Times*, 1 August 2006, A16. Blum is a Pulitzer Prize–winning science writer at the University of Wisconsin—Madison.

1. For biographical information on Simpson, see J[ohn] Duns, *Memoir of Sir James Y. Simpson, Bart.* (Edinburgh: Edmonston and Douglas, 1873).

2. W. O. Priestley and Horatio R. Storer, eds., *The Obstetric Memoirs and Contributions of James Y. Simpson, M.D. F.R.S.E.*, 2 vols. (Edinburgh: Adam and Charles Black, 1855), 1:608.

3. Ibid., 1:624.

4. Ibid., 1:623.

5. Ibid., 1:622.

6. Letter from Thomas Boodle to J. Y. Simpson, 14 January 1848, Royal College of Surgeons of Edinburgh, Surgeons' Hall Trust Collections, Archive RS S1 and RS S2, Papers of Sir James Young Simpson, J.Y.S. 224, http://www.rcsed.ac.uk/site/PID=42200414410/761/default.aspx (accessed 2 August 2007).

7. Letter from Rev. Robert Gayle to James Y. Simpson, 17 February 1848, Royal College of Surgeons of Edinburgh, Surgeons' Hall Trust Collections, Archive RS S1 and RS S2, Papers of Sir James Young Simpson, J.Y.S. 227, www.rcsed.ac.uk/site/PID=422004151120/761/default.aspx (accessed 2 August 2007).

8. A. D. Farr, "Religious Opposition to Obstetric Anaesthesia: A Myth?" *Annals of Science* 40 (1983): 166.

9. A. D. Farr, "Early Opposition to Obstetric Anaesthesia," *Anaesthesia* 35 (1980): 906.

10. Sylvia D. Hoffert, *Private Matters: American Attitudes toward Childbearing and Infant Nurture in the Urban North, 1800–1860* (Urbana: University of Illinois Press, 1989), 82, 87.

11. Walter Channing, *A Treatise on Etherization in Childbirth* (Boston: William D. Ticknor, 1848), "III. The Religious Objections to Etherization," 141–52, and "IV. The Moral Objection to Etherization," 152–56.

12. Ibid., 141, 142.

13. Ibid., 145, 149. Peppermint oil is a mild analgesic.

14. "Report of the Committee on Obstetrics," *Transactions of the American Medical Association* 1 (1848): 226, as quoted in Hoffert, *Private Matters,* 103, n. 122.

15. Hoffert, *Private Matters,* 90.

16. Donald Caton, *What a Blessing She Had Chloroform: The Medical and Social Response to the Pain of Childbirth from 1800 to the Present* (New Haven, Conn.: Yale University Press, 1999), 25. For a detailed discussion of the medical debates over chloroform, see A. J. Youngson, *The Scientific Revolution in Victorian Medicine* (New York: Holmes and Meier, 1979), chap. 3, "The Fight for Chloroform," 73–126.

17. For example, see Hoffert, *Private Matters,* 87.

18. Charles D. Meigs, *Obstetrics: The Science and the Art* (Philadelphia: Lea and Blanchard, 1849), 319.

19. Ibid., 325.

20. Martin S. Pernick, *A Calculus of Suffering: Pain, Professionalism, and Anesthesia in Nineteenth-Century America* (New York: Columbia University Press, 1985), 55.

21. White, *Warfare of Science with Theology,* 2:62–63; John William Draper, *History of the Conflict between Religion and Science* (New York: D. Appleton, 1874), 318–19.

22. Bertrand Russell, *Religion and Science* (New York: Oxford University Press, 1935), 105.

23. Thomas Dormandy, *The Worst of Evils: The Fight against Pain* (New Haven, Conn.: Yale University Press, 2006), 247.

24. "Pope Approves Pain-Easing Drugs, Even When Use Might Shorten Life," *Los Angeles Times,* 25 February 1957, 1, 12.

MYTH 15: THAT THE THEORY OF ORGANIC EVOLUTION IS BASED ON CIRCULAR REASONING

Epigraphs: Henry M. Morris, "Circular Reasoning in Evolutionary Biology," *Impact No. 48,* supplement to *Acts & Facts* 7 (June 1977). Jonathan Wells, *Icons of Evolution: Science or Myth?* (Washington, D.C.: Regnery Publishing, 2000), 61–62.

1. G. K. Chesterton, *The Everlasting Man,* in *The Collected Works of G. K. Chesterton,* vol. 2 (San Francisco: Ignatius Press, 1986), 266–67.

2. See also George McCready Price, *Evolutionary Geology and the New Catastrophism* (Mountain View, Calif.: Pacific Press Publishing Association, 1926), 9–43.

3. Henry M. Morris, "Science versus Scientism in Historical Geology," in *Scientific Studies in Special Creation,* ed. Walter E. Lammerts (Philadelphia: Presbyterian and Reformed Publishing Co., 1971), 116.

4. Ibid., 114.

5. John C. Whitcomb, Jr., and Henry M. Morris, *The Genesis Flood: The Biblical Record and Its Scientific Implications* (Philadelphia: Presbyterian and Reformed Publishing Co., 1961), 203–4; see also 132–35; 169–70; 203–6.

6. For further references pro and con, see Ronald L. Ecker, *Dictionary of Science and Creationism* (Buffalo, N.Y.: Prometheus Books, 1990), 101, 103.

7. For the religious affiliations of Wells and other creationists, see Ronald L. Numbers, *The Creationists: From Scientific Creationism to Intelligent Design,* exp. ed. (Cambridge, Mass.: Harvard University Press, 2006), 380–81 and *passim.*

8. Wells, *Icons of Evolution,* 63.

9. Ibid., 67–68. At the end Wells is quoting from Robert R. Sokal and Peter H. A. Sneath, *Principles of Numerical Taxonomy* (San Francisco: W. H. Freeman, 1963).

10. For example, David L. Hull, "Certainty and Circularity in Evolutionary Taxonomy," *Evolution* 21 (1967): 174–89.

11. Several quotations are given by Morris, "Circular Reasoning," and by Wells, *Icons of Evolution,* 63–66.

12. For a magisterial treatment of the relevant developments that highlight the continental contribution to the "reconstruction" of geological time, see Martin J. S. Rudwick, *Bursting the Limits of Time: The Reconstruction of Geohistory in the Age of Revolution* (Chicago: University of Chicago Press, 2005).

13. Taken from Nicolaas A. Rupke, *The Great Chain of History: William Buckland and the English School of Geology, 1814–1849* (Oxford: Clarendon Press, 1983), 111–29. See also Rachel Laudan, *From Mineralogy to Geology: The Foundations of a Science, 1650–1830* (Chicago: University of Chicago Press, 1987), 138–79.

14. Quoted in Rupke, *Great Chain of History*, 124.

15. Alexander von Humboldt, *A Geognostical Essay on the Superposition of Rocks, in Both Hemispheres* (London: Longman, Hurst, Rees, Orme, Brown, and Green, 1823), 3.

16. Humboldt appears to have subscribed to the theory of the autogenous generation of species. See Nicolaas A. Rupke, "Neither Creation nor Evolution: The Third Way in Mid-Nineteenth-Century Thinking about the Origin of Species," *Annals of the History and Philosophy of Biology* 10 (2005): 143–72, at 161–62. For more on the "third theory," see Rupke, "The Origin of Species from Linnaeus to Darwin," in *Aurora Torealis*, eds. Marco Beretta, Karl Grandin, and Svante Lindqvist (Sagamore Beach, Mass.: Science History Publications, 2008), 71–85.

17. On the history of geological progressionism, see Peter J. Bowler, *Fossils and Progress* (New York: Science History Publications, 1976).

18. Details are to be found in Nicolaas A. Rupke, *Richard Owen: Victorian Naturalist* (New Haven, Conn.: Yale University Press, 1994), 161–219.

19. A homologue was defined as "The same organ in different animals under every variety of form and function." An analogue, on the other hand, was "A part or organ in one animal which has the same function as another part or organ in a different animal." Richard Owen, *On the Archetype and Homologies of the Vertebrate Skeleton* (London: Richard and John E. Taylor, 1848), 7.

20. Nicolaas A. Rupke, "Richard Owen's Vertebrate Archetype," *Isis* 84 (1993): 231–51, at 249.

21. Rupke, "Richard Owen," 230–32 et seq.

22. Charles Darwin, *On the Origin of Species by Means of Natural Selection* (London: John Murray, 1859), 206, 435. On the incorporation of Owen's work in Darwinian evolutionary biology, see Adrian Desmond, *Archetypes and Ancestors: Palaeontology in Victorian London, 1850–1875* (London: Blonds and Briggs, 1982).

23. On the shaping of early representations of the stratigraphic column by Christian eschatologists, see Nicolaas A. Rupke, " 'The End of History' in the Early Picturing of Geological Time," *History of Science* 36 (1998): 61–90. The importance of progressionism in pre- and post-*Origin* geology and biology is discussed by Michael Ruse, *Monad to*

Man: The Concept of Progress in Evolutionary Biology (Cambridge, Mass.: Harvard University Press, 1996). The significance of Romantic holism for the growth of the homological research program is highlighted by Robert J. Richards, *The Romantic Conception of Life: Science and Philosophy in the Age of Goethe* (Chicago: University of Chicago Press, 2002).

MYTH 16: THAT EVOLUTION DESTROYED DARWIN'S FAITH IN CHRISTIANITY—UNTIL HE RECONVERTED ON HIS DEATHBED

Epigraphs: John D. Morris, "Did Darwin Renounce Evolution on His Deathbed?" *Back to Genesis,* no. 212 (August 2006): 1. L. R. Croft, *The Life and Death of Charles Darwin* (Chorley, Lancs.: Elmwood Books, 1989), 113–14. Malcolm Bowden, *True Science Agrees with the Bible* (Bromley, Kent: Sovereign Publications, 1998), 276.

1. Quoted in James R. Moore, "Of Love and Death: Why Darwin 'Gave Up Christianity,' " in *History, Humanity and Evolution: Essays for John C. Greene,* ed. James R. Moore (Cambridge: Cambridge University Press, 1989), 195–229, at 196.

2. Bowden, *True Science Agrees;* Christopher Chui, *Did God Use Evolution to "Create"? A Critique of Biological Evolution, Geological Evolution, and Astronomical Evolution* (Canoga Park, Calif.: Logos Publishers, 1993); Morris, "Did Darwin Renounce?"; Brian Eden, "Evolution v Creationism" [letter to the editor], *The Times* (London), 20 August 2003, responding to letters by Cyril Aydon on 18 August and Adrian Osmond on 12 August mentioning references to the renunciation in an evolution v. creationism debate.

3. R. D. Keynes, ed., *Charles Darwin's "Beagle" Diary* (Cambridge: Cambridge University Press, 1988), 444.

4. Paul H. Barrett et al., eds., *Charles Darwin's Notebooks, 1836–1844: Geology, Transmutation of Species, Metaphysical Enquiries* (Cambridge: Cambridge University Press, 1987), 300.

5. Ibid., 263.

6. Ibid., 343.

7. Francis Darwin, ed., *The Foundations of the Origin of Species: Two Essays Written in 1842 and 1844* (Cambridge: University Press, 1909), 51–52.

8. Charles Darwin, *On the Origin of Species by Means of Natural Selection, or the Preservation of Favoured Races in the Struggle for Life* (London: John Murray, 1859), 490.

9. Charles Darwin, *The Descent of Man, and Selection in Relation to Sex*, 2 vols. (London: John Murray, 1871), 2:394, 2:396.

10. Nora Barlow, ed., *The Autobiography of Charles Darwin, 1809–1882, with Original Omissions Restored* (London: Collins, 1958), 85–87.

11. Quoted in James R. Moore, *The Darwin Legend* (Grand Rapids, Mich.: Baker, 1994), 125, 131.

12. Quoted in ibid., 92–93.

13. Referenced in ibid., 95–97.

14. Emma Darwin's diary, 7 November 1881, Darwin Archive, Cambridge University Library, DAR 242:45.

15. Quoted in Moore, *Darwin Legend*, 52.

MYTH 17: THAT HUXLEY DEFEATED WILBERFORCE IN THEIR DEBATE OVER EVOLUTION AND RELIGION

Epigraph: John H. Lienhard, "Soapy Sam and Huxley," Episode No. 1371, 1998, "Engines of Our Ingenuity," broadcast available at http://www.uh.edu/engines/epi1371.htm (accessed 19 June 2008).

1. William Irvine, *Apes, Angels and Victorians: A Joint Biography of Darwin and Huxley* (London: Weidenfeld and Nicolson, 1956), 5–6.

2. Andrew D. White, *A History of the Warfare of Science with Theology in Christendom*, 2 vols. (New York: D. Appleton, 1896), 1:70–71.

3. Sheridan Gilley, "The Huxley-Wilberforce Debate: A Reconsideration," in *Religion and Humanism*, ed. Keith Robbins, *Studies in Church History* 17 (Oxford: Blackwell, 1981): 325–40, at 325.

4. M. M. Woolfson, "Conform or Think?" *Astronomy & Geophysics* 45, 5 (2004): 5.8.

5. Frank James, "On Wilberforce and Huxley," *Astronomy & Geophysics* 46 (February 2005): 1.9.

6. In addition to the sources itemized below, see also Stephen Jay Gould, "Knight Takes Bishop?" *Natural History* 95 (May 1986): 18–33.

7. Leonard Huxley, *Life and Letters of Thomas Henry Huxley*, 2 vols. (London: Macmillan, 1900), 1:181.

8. This point, with accompanying extensive quotations, is made in J. R. Lucas, "Wilberforce and Huxley: A Legendary Encounter," *Historical Journal* 22 (1979): 313–30; and in Josef L. Altholz, "The Huxley-Wilberforce Debate Revisited," *Journal of the History of Medicine and Allied Sciences* 35 (1980): 313–16.

9. See Frank A. J. L. James, "An 'Open Clash between Science and the Church'? Wilberforce, Huxley and Hooker on Darwin at the British Association, Oxford, 1860," in *Science and Beliefs: From Natural Philosophy to Natural Science, 1700–1900,* ed. David M. Knight and Matthew D. Eddy (Aldershot: Ashgate, 2005), 171–93.

10. Extracts quoted in Keith Stewart Thomson, "Huxley, Wilberforce and the Oxford Museum," *American Scientist* 88 (May–June 2000): 210–13.

11. Quoted in ibid.

12. See John Hedley Brooke, "The Wilberforce-Huxley Debate: Why Did It Happen?" *Science and Christian Belief* 13 (2001): 127–41.

13. This point is made in Sheridan Gilley and Ann Loades, "Thomas Henry Huxley: The War between Science and Religion," *Journal of Religion* 61 (1981): 285–308.

14. Quoted in Lucas, "Wilberforce and Huxley," 318.

15. Quoted in Brooke, "Wilberforce-Huxley Debate," 139.

16. Quoted in Adrian Desmond and James Moore, *Darwin* (London: Michael Joseph, 1991), 487. Kenneth J. Howell uses Wilberforce's philosophical concerns as a foil for reflecting on Pope John Paul II's perspective on science and religion. See Howell, "Did the Bulldog Bite the Bishop? An Anglican Bishop, an Agnostic Scientist, and a Roman Pontiff," *Logos: A Journal of Catholic Thought and Culture* 6 (Summer 2003): 41–67.

17. Adrian Desmond, *Huxley: The Devil's Disciple* (London: Penguin, 1994).

18. See Frank Miller Turner, "The Victorian Conflict between Science and Religion: A Professional Dimension," *Isis* 69 (1978): 356–76.

19. Quoted in Desmond and Moore, *Darwin,* 517.

20. Quoted in Lucas, "Wilberforce and Huxley," 327.

21. James Y. Simpson, *Landmarks in the Struggle between Science and Religion* (London: Hodder and Stoughton, 1925), 192. This Simpson was the great-nephew of the James Young Simpson discussed in Myth 14.

22. Janet Browne, *Charles Darwin: The Power of Place* (London: Jonathan Cape, 2002), 122.

23. Paul White, *Thomas Huxley: Making the "Man of Science"* (Cambridge: Cambridge University Press, 2003), 65.

24. See Lynn A. Phelps and Edwin Cohen, "The Wilberforce-Huxley Debate," *Western Speech* 37 (1973): 57–60; and J. Vernon Jensen, "Return to the Wilberforce-Huxley Debate," *British Journal for the History of Science* 21 (1988): 161–79.

25. William MacIlwaine, Address, *Proceedings of the Belfast Naturalists' Field Club* (1874–75): 81–99, at 82; *McComb's Presbyterian Almanack, and Christian Remembrancer for 1875* (Belfast: James Cleeland, 1875), 84; John Tyndall, *Fragments of Science*, 6th ed. (New York: D. Appleton, 1889), 472–534, on 530.

26. John William Draper, *History of the Conflict between Religion and Science* (New York: D. Appleton, 1874).

MYTH 18: THAT DARWIN DESTROYED NATURAL THEOLOGY

I would be remiss if I did not express appreciation to members of the "myth-busting" conference devoted to discussion of the essays in this collection for their helpful comments. I owe a special debt of gratitude to Ron Numbers for his usual excellent advice and editorial assistance and to my wife, Sharon (ILYS), for her ongoing love and support of me and my work.

Epigraphs: Ernst Mayr, *The Growth of Biological Thought* (Cambridge, Mass.: Harvard University Press, 1982), 515. T. M. Heyck, *The Transformation of Intellectual Life in Victorian England* (Chicago: Lyceum Books, 1982), 85. Richard Dawkins, *The Blind Watchmaker: Why the Evidence of Evolution Reveals a Universe without Design* (New York: W. W. Norton, 1986), 6.

1. John Hedley Brooke, "Indications of a Creator: Whewell as Apologist and Priest," in *William Whewell: A Composite Portrait*, ed. Menachem Fisch and Simon Schaffer (Oxford: Clarendon Press, 1991), 149–73; John Hedley Brooke, "The Natural Theology of the Geologists: Some Theological Strata," in *Images of the Earth: Essays in the History of the Environmental Sciences*, ed. L. J. Jordanova and Roy S. Porter (Chalfont St. Giles: British Society for the History of Science, 1979), 39–41, passim; Jonathan R. Topham, "Science, Natural Theology, and the Practice of Christian Piety in Early-Nineteenth-Century Religious Magazines," in *Science Serialized: Representation of the Sciences in Nineteenth-Century Periodicals*, ed. Geoffrey Cantor and Sally Shuttleworth (Cambridge, Mass.: MIT Press, 2004), 37–66; and Aileen Fyfe, *Science and Salvation: Evangelical Popular Science Publishing in Victorian Britain* (Chicago: University of Chicago Press, 2004), 6–7.

2. Jon H. Roberts, *Darwinism and the Divine in America: Protestant Intellectuals and Organic Evolution, 1859–1900* (Madison: University of Wisconsin Press, 1988), 10–11.

3. Thomas H. Huxley, "Criticisms on 'The Origin of Species'" [1864], *Darwiniana*, in *Collected Essays by T. H. Huxley*, vol. 2 (1893; New York: Greenwood Press, 1968), 82.

4. Daniel R. Goodwin, "The Antiquity of Man," *American Presbyterian and Theological Review,* n.s., 2 (1864): 259. See also Henry M. Harman, "Natural Theology," *Methodist Review,* 4th ser., 15 (1863): 182–90.

5. James T. Bixby, "The Argument from Design in the Light of Modern Science," *Unitarian Review and Religious Magazine,* 8 (1877): 21–23; Asa Gray, "Natural Selection Not Inconsistent with Natural Theology" [1860], *Darwiniana: Essays and Reviews Pertaining to Darwinism,* ed. A. Hunter Dupree (1876; Cambridge, Mass.: Harvard University Press, 1963), 121–22, 129; Anonymous, "Current Skepticism—The Scientific Basis of Faith," *Methodist Review,* 5th ser., 8 (1892): 953.

6. Charles Darwin, *The Variation of Animals and Plants under Domestication,* 2 vols. (London: John Murray, 1868), 2:430–32.

7. Asa Gray, *Natural Science and Religion: Two Lectures Delivered to the Theological School of Yale College* (New York: Charles Scribner's Sons, 1880), 47; Andrew P. Peabody, "Science and Revelation," *Princeton Review,* 4th ser., 54th yr. (1878): 766.

8. James McCosh, *The Religious Aspect of Evolution,* rev. ed (New York: Charles Scribner's Sons, 1890), 70; Bixby, "Argument from Design," 4–5, on 5; F. A. Mansfield, "Teleology, Old and New," *New Englander,* n.s., 7 (1884): 220. See also J. Lewis Diman, *The Theistic Argument as Affected by Recent Theories: A Course of Lectures Delivered at the Lowell Institute in Boston* (Boston: Houghton, Mifflin, 1881), 178–79.

9. Diman, *Theistic Argument,* 165, 167.

10. McCosh, *Religious Aspect of Evolution,* 7.

11. James Iverach, quoted in David W. Bebbington, "Science and Evangelical Theology in Britain from Wesley to Orr," in *Evangelicals and Science in Historical Perspective,* ed. David N. Livingstone et al. (New York: Oxford University Press, 1999), 132.

12. Bixby, "Argument from Design," 23–26; George F. Wright, "Recent Works Bearing on the Relation of Science to Religion: No. IV— Concerning the True Doctrine of Final Cause or Design in Nature," *Bibliotheca Sacra* 34 (1877): 358.

13. D. B. Purinton, *Christian Theism: Its Claims and Sanctions* (New York: G. P. Putnam's Sons, 1889), 31; George Harris, *Moral Evolution* (Boston: Houghton, Mifflin and Company, 1896), 185. See also George P. Fisher, "Materialism and the Pulpit," *Princeton Review,* 4th ser., 54th yr. (1878): 207–9.

14. J. W. Chadwick, "The Basis of Religion," *Unitarian Review and Religious Magazine* 26 (1886): 255–56. See also [G. F.] W[right], Re-

view of *The Evolution of Man,* by Ernst Haeckel, *Bibliotheca Sacra* 36 (1879): 784; William North Rice, "Evolution" [1890], in *Twenty-Five Years of Scientific Progress and Other Essays* (New York: Thomas Y. Crowell, 1894), 86.

15. Lewis F. Stearns, "Reconstruction in Theology," *New Englander,* n.s., 5 (1882): 86.

16. L. E. Hicks, *A Critique of Design-Arguments: A Historical Review and Free Examination of the Methods of Reasoning in Natural Theology* (New York: Charles Scribner's Sons, 1883), 331.

17. Joseph LeConte, *Evolution and Its Relation to Religious Thought* (New York: D. Appleton, 1888), 283.

18. Bernard Lightman, "Victorian Sciences and Religions: Discordant Harmonies," *Osiris,* 2d ser., 16 (2001): 355–62. See also Bernard Lightman, *Victorian Popularizers of Science: Designing Nature for New Audiences* (Chicago: University of Chicago Press, 2007).

19. For the persistence of the argument from design among philosophers see, for example, F. R. Tennant, *Philosophical Theology,* 2 vols. (Cambridge: Cambridge University Press, 1930). Useful treatments of American neo-Scholasticism include Philip Gleason, *Contending with Modernity: Catholic Higher Education in the Twentieth Century* (New York: Oxford University Press, 1995), esp. 114–23, and William M. Halsey, *The Survival of American Innocence: Catholicism in an Era of Disillusionment, 1920–1940* (Notre Dame, Ind.: University of Notre Dame Press, 1980), 138–68.

20. Peter J. Bowler, *Reconciling Science and Religion: The Debate in Early-Twentieth-Century Britain* (Chicago: University of Chicago Press, 2001), 122–59.

21. James Jeans, *The Mysterious Universe,* rev. ed. (1932; New York: Macmillan Company, 1933), 186, 175, 17–53, 165, 186–87. See also Arthur Eddington, *The Nature of the Physical World* (1928; Ann Arbor: University of Michigan Press, 1958), 276–82.

22. Harry Emerson Fosdick, *Adventurous Religion and Other Essays* (New York: Cornwall Press, 1926), 212–13.

23. For a more extensive discussion of the issues broached in this paragraph, see Jon H. Roberts, "Science and Religion," in *Wrestling with Nature: From Omens to Science,* ed. Peter Harrison, Ronald L. Numbers, and Michael Shank (Chicago: University of Chicago Press, forthcoming).

24. Nancey Murphy, *Theology in the Age of Scientific Reasoning* (Ithaca, N.Y.: Cornell University Press, 1990), 18.

25. John Polkinghorne, *One World: The Interaction of Science and Theology* (Princeton, N.J.: Princeton University Press, 1986), 81.

26. William A. Dembski, "Introduction," in *Mere Creation: Science, Faith and Intelligent Design*, ed. William A. Dembski (Downers Grove, Ill.: InterVarsity Press, 1998), 16; Polkinghorne, *One World*, 78.

27. Polkinghorne, *One World*, 79; John Polkinghorne, *The Faith of a Physicist: Reflections of a Bottom-Up Thinker; The Gifford Lectures for 1993–4* (Princeton, N.J.: Princeton University Press, 1994), 43; John Polkinghorne, *Belief in God in an Age of Science* (New Haven, Conn.: Yale University Press, 1998), 10–11.

28. Arthur Peacocke, *Theology for a Scientific Age: Being and Becoming—Natural, Divine and Human*, rev. ed. (Minneapolis: Fortress Press, 1993), 134.

MYTH 19: THAT DARWIN AND HAECKEL WERE COMPLICIT IN NAZI BIOLOGY

Epigraphs: Stephen Jay Gould, *Ontogeny and Phylogeny* (Cambridge, Mass.: Harvard University Press, 1977), 77. Richard Weikart, *From Darwin to Hitler: Evolutionary Ethics, Eugenics, and Racism in Germany* (New York: Palgrave Macmillan, 2004), 6.

1. Gould, *Ontogeny and Phylogeny*, 77.

2. Charles Darwin to Ernst Haeckel, 9 March 1864, *The Correspondence of Charles Darwin*, ed. Frederick Burkhardt et al., 15 vols. (Cambridge: Cambridge University Press, 1985–),12:63.

3. Charles Darwin, *The Descent of Man, and Selection in Relation to Sex*, 2 vols. (London: John Murray, 1871), 1:4.

4. I have argued this at greater length in my *The Tragic Sense of Life: Ernst Haeckel and the Struggle over Evolutionary Thought* (Chicago: University of Chicago Press, 2008), chap. 5.

5. Weikart, *From Darwin to Hitler*.

6. Charles Darwin, *On the Origin of Species* (London: John Murray, 1859), 489.

7. Ibid., 450.

8. Georges Cuvier, *Le Régne animal*, 2d ed., 5 vols. (Paris: Deterville Libraire, 1829–1830), 1:80.

9. Darwin, *Descent of Man*, 1:235.

10. See Charles Darwin to Asa Gray, 19 April 1865, in *Correspondence of Charles Darwin*, 13:126.

11. Daniel Gasman, *Haeckel's Monism and the Birth of Fascist Ideology* (New York: Peter Lang, 1998), 26.

NOTES TO PAGES 175-180

12. Hermann Bahr, *Der Antisemitismus* (Berlin: S. Fischer, 1894), 69.

13. Hirschfeld to Haeckel, 17 December 1914, in the Correspondence of Ernst Haeckel, Ernst-Haeckel-Haus, Jena.

14. I discuss Haeckel's alleged anti-Semitism at greater length in my *Tragic Sense of Life*.

15. Jürgen Sandmann, *Der Bruch mit der humanitären Tradition: Die Biologisierung der Ethik bei Ernst Haeckel und anderen Darwinisten seiner Zeit* (Stuttgart: Gustav Fischer, 1990).

16. Darwin, *Descent of Man*, 1:161–67.

17. Ernst Haeckel, *Der Monismus als Band zwischen Religion und Wissenschaft* (Bonn: Emil Strauss, 1892), 29.

18. Ernst Haeckel, *Natürliche Schöpfungsgeschichte*, 2d ed. (Berlin: Georg Reimer, 1870), 156.

19. Nicolaas Rupke, *Alexander von Humboldt: A Metabiography* (Frankfurt am Main: Peter Lang, 2005), 81–104.

20. See, for instance, Heinz Brücher, *Ernst Haeckels Bluts- und Geistes-Erbe: Eine kulturbiologische Monographie* (Munich: Lehmanns Verlag, 1936).

21. Günther Hecht, "Biologie und Nationalsozialismus," *Zeitschrift für die Gesamte Naturwissenschaft* 3 (1937–1938): 280–90, at 285. This journal bore the subtitle *Organ of the Reich's Section Natural Science of the Reich's Students Administration.*

22. Kurt Hildebrandt, "Die Bedeutung der Abstammungslehre für die Weltanschauung," *Zeitschrift für die Gesamte Naturwissenschaft* 3 (1937–1938): 15–34, at 17.

23. "Richtilinien für die Bestandsprüfung in den Volksbüchereien Sachsens," *Die Bücherei* 2 (1935): 279–80.

MYTH 20: THAT THE SCOPES TRIAL ENDED IN DEFEAT FOR ANTIEVOLUTIONISM

Epigraph: William E. Leuchtenburg, *The Perils of Prosperity, 1914–1932* (Chicago: University of Chicago Press, 1958), 223.

1. For the full story, see Edward J. Larson, *Summer for the Gods: The Scopes Trial and America's Continuing Debate over Science and Religion* (New York: Basic Books, 1997).

2. Frederick Lewis Allen, *Only Yesterday: An Informal History of the Nineteen-Twenties* (1931; New York: Harper and Row, 1964), 171.

3. Thomas A. Bailey, *The American Pageant: A History of the Republic,* 2d ed. (Boston: Heath, 1961), 795.

4. Samuel Eliot Morison, Henry Steele Commager, and William E. Leuchtenburg, *History of the American Republic,* 6th ed., 2 vols. (New York: Oxford University Press, 1969), 2:436.

5. Richard Hofstadter et al., *The American Republic,* 2d ed., 2 vols. (Englewood Cliffs, N.J.: Prentice-Hall, 1970), 2:389.

6. Richard Hofstadter, *Anti-Intellectualism in American Life* (New York: Random House, 1962), 129.

7. Jerome Lawrence and Robert E. Lee, *Inherit the Wind* (New York: Bantam, 1960), 58–59.

8. Ibid., 7, 64. *Inherit the Wind* uses sound-alike pseudonyms in naming the various participants in the Scopes trial. To avoid unnecessary confusion, the real names of persons are used in this article even when discussing actions of their characters in *Inherit the Wind.*

9. Lawrence and Lee, *Inherit the Wind,* 4 (stage directions regarding minister's daughter).

10. *World's Most Famous Court Case: Tennessee Evolution Case* (Dayton, Tenn.: Bryan College, 1990), 288 (reprint of trial transcript).

11. Ralph Perry, "Added Thrill Given Dayton," *Nashville Banner,* 21 July 1925, 2.

12. Clarence Darrow, *The Story of My Life* (New York: Grosset, 1932), 267.

13. *World's Most Famous Court Case,* 285.

14. Ibid., 302.

15. Ibid.

16. Ibid., 304.

17. Lawrence and Lee, *Inherit the Wind,* 63.

18. Ibid., 85.

19. Ibid., 91.

20. Ibid., 103.

21. Ibid.

22. Ibid., 109.

23. Ibid., 113.

24. Ibid., 112–15.

25. H. L. Mencken, "Editorial," *American Mercury* 6 (1925): 159.

MYTH 21: THAT EINSTEIN BELIEVED IN A PERSONAL GOD

Epigraphs: Charles Krauthammer, "Phony Theory, False Conflict," *Washington Post,* 18 November 2005, A23; this op-ed piece may also be accessed at http://www.washingtonpost.com. Stephen Caesar, "Investigating Origins: Einstein and Intelligent Design," accessed at http://familyac tionorganization.wordpress.com/2007/08/17/investigatingoriginseinstein-and-intelligent-design/ (accessed 8 September 2008).

1. Several versions of this story exist, with different people playing the part of the student. A common form of the Einstein variant can be seen at http://www.snopes.com/religion/einstein.asp.

2. Albert Einstein, "Science and Religion," in *Ideas and Opinions* (New York: Crown Publishers, 1954), 41–49, at 46–47 and 48.

3. Max Jammer, *Einstein and Religion* (Princeton, N.J.: Princeton University Press, 1999), 122–23. Jammer is the best single source for Einstein's views on religion and religious uses of Einstein's ideas.

4. Albert Einstein, "Religion and Science," in *Ideas and Opinions*, 6–40, at 37.

5. Ibid., 38; Einstein, "Science and Religion," 48; Jammer, *Einstein and Religion*, 149.

6. Helen Dukas, ed., *Albert Einstein: The Human Side* (Princeton, N.J.: Princeton University Press, 1979), 39, 95.

7. Jammer, *Einstein and Religion*, 123.

8. Abraham Pais, *'Subtle is the Lord . . .': The Science and the Life of Albert Einstein* (Oxford: Clarendon Press, 1982), vi; Jammer, *Einstein and Religion*, 124.

9. Dukas, *Albert Einstein*, 43; Albert Einstein, *Autobiographical Notes*, ed. P. A. Schilpp (La Salle, Ill.: Open Court Press, 1979), 3–5.

10. Einstein, "Science and Religion," 44–45; Einstein to Maurice Solovine, 1 January 1951, in Albert Einstein, *Letters to Solovine*, trans. Wade Baskin (New York: Philosophical Library, 1987), 119; Gerald Holton, "Einstein's Third Paradise," *Dædalus* 132 (2003): 26–34, at 31.

11. Albert Einstein, "The Religious Spirit of Science," in *Ideas and Opinions*, 40.

12. Einstein, "Religion and Science," 38; Einstein, "Religious Spirit of Science," 40.

13. Einstein to Phyllis, 24 January 1936, in *Dear Professor Einstein*, ed. Alice Calaprice (Amherst, N.Y.: Prometheus Books, 2002), 128–29; Einstein, "Science and Religion," 46.

14. A. S. Eddington, *Philosophy of Physical Science* (Cambridge: Cambridge University Press, 1939), 7. Not everyone agreed that relativity had no religious significance. Eddington, who reported Einstein's response, pointed out that natural selection was also a purely scientific theory but clearly had important religious implications. For more on

how relativity has functioned in religious contexts, see Jammer, *Einstein and Religion*, 153–266, and Matthew Stanley, *Practical Mystic: Religion, Science, and A. S. Eddington* (Chicago: University of Chicago Press, 2007), 153–247. It has been speculated that Einstein's commitment to causality in quantum mechanics and a static universe in cosmology can be linked to his Spinozistic religion (see Jammer, *Einstein and Religion*, 62–63).

15. Einstein, "Religion and Science," 40.

16. Jammer, *Einstein and Religion*, 48; and Einstein to Solovine, 30 March 1952, in *Letters to Solovine*, 131–33.

17. Jammer, *Einstein and Religion*, 50–51; Albert Einstein, "The World as I See It," in *Ideas and Opinions*, 11; and Dukas, *Albert Einstein*, 81.

18. Homer W. Smith, *Man and His Gods* (Boston: Little, Brown and Co., 1953), ix; Einstein, *Autobiographical Notes*, 3.

19. Albrecht Fölsing, *Albert Einstein: A Biography* (New York: Viking, 1997), 41, 273; Jammer, *Einstein and Religion*, 25–27; Albert Einstein, "A Letter to Dr. Hellpach, Minister of State," in *Ideas and Opinions*, 171–72, on 171; Jammer, *Einstein and Religion*, 59–60; Albert Einstein, "Is There a Jewish Point of View?" in *Ideas and Opinions*, 185–87, at 186.

MYTH 22: THAT QUANTUM PHYSICS DEMONSTRATED THE DOCTRINE OF FREE WILL

Epigraphs: Waldemar Kaempffert, "St. Louis Conference Considers Some Basic Problems in the Thinking of Modern Men," *New York Times*, 31 October 1954, E9. Denyse O'Leary, "The ID Report," 23 September 2006, http://www.arn.org/blogs/index.php/2/2006/09/23/lstrong glemgquantum_mechanics_l_emg_does (accessed 3 March 2007).

1. Amit Goswami, with Richard Reed and Maggie Goswami, *The Self-Aware Universe* (New York: Putnam, 1993), xvi.

2. For later elaborations on the many-worlds interpretation, see Bryce S. DeWitt and Neill Graham, eds., *The Many-Worlds Interpretation of Quantum Mechanics* (Princeton, N.J.: Princeton University Press, 1973); on Bohm's development of his ideas, see David Bohm, *Wholeness and the Implicate Order* (London: Routledge and Kegan Paul, 1981).

3. On Eddington's views regarding science and religion, see Allen H. Batten, "A Most Rare Vision: Eddington's Thinking on the Relation between Science and Religion," *Quarterly Journal of the Royal*

Astronomical Society 35 (1994): 249–70; and Matthew Stanley, *Practical Mystic: Religion, Science, and A. S. Eddington* (Chicago: University of Chicago Press, 2007).

4. Allen H. Batten, "What Eddington Did *Not* Say," *Isis* 94 (2003): 658.

5. "Sees Deity Ruling World of Chance," *New York Times*, 27 March 1931, 27.

6. See, for instance, William Savery, "Chance and Cosmogony," *Proceedings and Addresses of the American Philosophical Association* 5 (1931): 176; and Charles G. Darwin, *The New Conceptions of Matter* (London: G. Bell and Sons, 1931), 118.

7. "The Scientific Mind," *New York Times*, 15 February 1931, 67.

8. William P. Montague, "Beyond Physics," *Saturday Review of Literature*, 23 March 1929, 801.

9. For a brief mention of Hsieh's book, see Wolfgang Pauli, "Modern Examples of 'Background Physics,' " in *Atom and Archetype*, ed. C. A. Meier, trans. David Roscoe (Princeton, N.J.: Princeton University Press, 2001), 188.

10. The Dalai Lama, *The Universe in a Single Atom* (New York: Morgan Road Books, 2005), 58.

11. Goswami, *Self-Aware Universe*, 141.

12. Henry P. Stapp, *Mind, Matter, and Quantum Physics* (Berlin: Springer-Verlag, 1993), 20–21.

13. Apryl Jensen, "Quantum Physics and You—How the Quantum Science of the Unseen Can Transform Your Life!" at http://www.creatingconsciously.com/quantumphysics.html (accessed 5 July 2007).

14. See, for instance, Victor J. Stenger, *The Unconscious Quantum* (Amherst, N.Y.: Prometheus Books, 1993); and Stenger, "Quantum Quackery," *Skeptical Inquirer* 21 (1997): 37–40.

15. Robert J. Russell, "Theology and Quantum Theory," in *Physics, Philosophy, and Theology: A Common Quest for Understanding*, ed. Robert J. Russell, W. R. Stoeger, and G. V. Coyne (Notre Dame, Ind.: University of Notre Dame Press, 1988). See also Gregory R. Peterson, "God, Determinism, and Action: Perspectives from Physics," *Zygon* 35 (2000): 882.

16. See for instance Thomas J. McFarlane, ed., *Einstein and Buddha: The Parallel Sayings* (Berkeley, Calif.: Seastone, 2002), which includes sayings not only by Einstein but also by Bohr, Heisenberg, Pauli, Max Planck, Erwin Schrödinger, and other twentieth-century physicists.

17. Abraham Pais, *Niels Bohr's Times* (Oxford: Clarendon Press, 1991), 24, 310–11, 420–425.

18. Werner Heisenberg, "Science and Religion," in *Physics and Beyond* (New York: Harper and Row, 1971).

19. Pais, *Niels Bohr's Times*, 440–41.

20. David Cassidy, *Uncertainty: The Life and Science of Werner Heisenberg* (New York: W. H. Freeman and Company, 1992), 46–48; Werner Heisenberg, *Physics and Philosophy* (New York: Harper and Brothers, 1958), 128.

21. Heisenberg, *Physics and Philosophy*, 58; Heisenberg, "If Science Is Conscious of Its Limits . . . ," in *The Physicist's Conception of Nature* (New York: Harcourt and Brace, 1955), reprinted in *Quantum Questions*, ed. Ken Wilbur (Boston: New Science Library, 1985), 73.

22. Heisenberg, "Science and Religion." For a list of the other areas, from biology to psychology, in which Bohr was willing to apply complementarity beginning in the late 1920s, see Pais, *Niels Bohr's Times*, 438–47.

23. Heisenberg, "Science and Religion." Ken Wilbur has compiled a variety of writings by the founders of quantum theory to demonstrate their views on the relationship between science and religion. See Wilbur, *Quantum Questions*.

24. Heisenberg, *Physics and Philosophy*, 137.

25. Ibid., 58.

26. Stapp, *Mind, Matter, and Quantum Physics*, 129–30.

27. For Pauli and Jung's collected correspondence, see Meier, *Atom and Archetype*.

28. Heisenberg, *Physics and Philosophy*, 187.

29. Ibid., 202.

30. Dalai Lama, *Universe in a Single Atom*, 65.

MYTH 23: THAT "INTELLIGENT DESIGN" REPRESENTS A SCIENTIFIC CHALLENGE TO EVOLUTION

Epigraphs: Michael J. Behe, *Darwin's Black Box: The Biochemical Challenge to Evolution* (New York: Free Press, 1996), 232–33. Peter Baker and Peter Slevin, "Bush Remarks on 'Intelligent Design' Theory Fuel Debate," *Washington Post*, 3 August 2005, A1. William Dembski, "Why President Bush Got It Right about Intelligent Design," 4 October 2005, http://www/designinference.com/documents/2005.08.Commending_President_Bush.pdf (accessed 23 June 2008).

1. Percival Davis and Dean H. Kenyon, *Of Pandas and People: The Central Question of Biological Origins*, 2d ed. (1989; Dallas:

Haughton Publishing, 1993), 14, 160–61; Robert T. Pennock, *Tower of Babel: The Evidence against the New Creationism* (Cambridge, Mass.: MIT Press, 1999), 276 (politically correct). The history of the *Pandas* book appears in Barbara Forrest's testimony in *Tammy Kitzmiller, et al. v. Dover Area School District*, et al. 5 October 2005, a transcript of which can be found at the National Center for Science Education's web site, www2.ncseweb.org. For a brief history of intelligent design, see Ronald L. Numbers, *The Creationists: From Scientific Creationism to Intelligent Design*, exp. ed. (Cambridge, Mass.: Harvard University Press, 2006), chap. 17, "Intelligent Design."

2. Phillip E. Johnson, *Darwin on Trial* (Downers Grove, Ill.: InterVarsity Press, 1991). This and the following paragraphs are extracted from Michael Ruse, *The Evolution-Creation Struggle* (Cambridge, Mass.: Harvard University Press, 2005), 250–55.

3. Phillip E. Johnson, "Foreword," in *The Creation Hypothesis: Scientific Evidence for an Intelligent Designer*, ed. J. P. Moreland (Downers Grove, Ill.: InterVarsity Press, 1994), 7–8; Phillip E. Johnson, *Defeating Darwinism by Opening Minds* (Downers Grove, Ill.: InterVarsity Press, 1997), 92 (log). For a critical history of "the wedge," see Barbara Forrest and Paul R. Gross, *Creationism's Trojan Horse: The Wedge of Intelligent Design* (New York: Oxford University Press, 2004), esp. 3–47.

4. Behe, *Darwin's Black Box*, 39.

5. Ibid., 70.

6. William A. Dembski, "The Explanatory Filter: A Three-Part Filter for Understanding How To Separate and Identify Cause from Intelligent Design," an excerpt from a paper presented at the 1996 "Mere Creation" conference, originally titled "Redesigning Science," available at http://www.arn.org/docs/dembski/wd_explfilter.htm. See also William A. Dembski, *The Design Inference: Eliminating Chance through Small Probabilities* (Cambridge: Cambridge University Press, 1998); and William A. Dembski, *Mere Creation: Science, Faith and Intelligent Design* (Downers Grove, Ill.: InterVarsity Press, 1998).

7. William A. Dembski, *Intelligent Design: The Bridge between Science and Theology* (Downers Grove, Ill.: InterVarsity Press, 1999), 206; William A. Dembski, "Signs of Intelligence: A Primer on the Discernment of Intelligent Design," in *Signs of Intelligence: Understanding Intelligent Design*, ed. William A. Dembski and James M. Kusiner (Grand Rapids, Mich.: Brazos, 2001), 192. The Behe comment appears on Larry Arnhart's blog, "Darwinian Conservatism," 7 September 2006, at

http://darwinianconservatism.blogspot.com/2006/09/has-anyone-seen-evolution.html (accessed 1 July 2008).

8. Michael Behe to the author, June 2003, quoted in Ruse, *Evolution-Creation Struggle*, 256; William Dembski, "Signs of Intelligence: A Primer on the Discernment of Intelligent Design" *Touchstone* 12 (July/August 1999): 76–84, at 84; Johnson, *Darwin on Trial*, 157–58.

9. David K. Webb, Letter to the Editor, *Origins & Design* 17 (Spring 1996): 5.

10. Michael Ruse, "Creation-Science Is Not Science," in *Creationism, Science, and the Law: The Arkansas Case*, ed. Marcel Chotkowski La Follette (Cambridge, Mass.: MIT Press, 1983), 150–60.

11. William Whewell, *History of the Inductive Sciences: From the Earliest to the Present Time*, 3d ed., 2 vols. (1837; New York: D. Appleton, 1858), 2:573; Charles Darwin, *The Variation of Animals and Plants Under Domestication*, 2 vols. (London: John Murray, 1868), 2:516. For a history of methodological naturalism, see Ronald L. Numbers, "Science without God: Natural Laws and Christian Beliefs," in *When Science and Christianity Meet*, ed. David C. Lindberg and Ronald L. Numbers (Chicago: University of Chicago Press, 2003), 265–85.

12. Alvin Plantinga, "Methodological Naturalism?" *Perspectives on Science and Christian Faith* 49 (September 1997): 143–54, at 152–53. See also Alvin Plantinga, "An Evolutionary Argument against Naturalism," *Logos* 12 (1991): 27–49.

13. Dennis Overbye, "Philosophers Notwithstanding, Kansas School Board Redefines Science," *New York Times*, 15 November 2005, D3; Charles Krauthammer, "Phony Theory, False Conflict: 'Intelligent Design' Foolishly Pits Evolution against Faith," *Washington Post*, 18 November 2005, A23.

14. Complaint filed by the plaintiffs in *Tammy Kitzmiller, et al. v. Dover Area School District et al.* See also Nicholas J. Matzke, "Design on Trial in Dover, Pennsylvania," *NCSE Reports* 24 (September/October 2004): 4–9; Neela Banerjee, "School Board Sued on Mandate for Alternative to Evolution," *New York Times*, 15 December 2004, A25; Laurie Goodstein, "A Web of Faith, Law and Science in Evolution Suit," ibid., 26 September 2005, A1, A14; Laurie Goodstein, "Evolution Lawsuit Opens with Broadside against Intelligent Design," ibid., 27 September 2005, A17; and Constance Holden, "ID Goes on Trial This Month in Pennsylvania School Case," *Science* 309 (2005): 1796.

15. Steve William Fuller, "Rebuttal of Dover Expert Reports," 13 May 2005; testimony of Steve Fuller, *Tammy Kitzmiller, et al. v. Dover*

Area School District, et al. 24 October 2005. Fuller had earlier expressed his support for intelligent design in a letter to the editor, *Chronicle of Higher Education,* 1 February 2002, B4, B17.

16. John E. Jones III, "Memorandum Opinion," 20 December 2005, found on the NCSE's web site. See also Jeffrey Mervis, "Judge Jones Defines Science—and Why Intelligent Design Isn't," *Science* 311 (2006): 34.

MYTH 24: THAT CREATIONISM IS A UNIQUELY AMERICAN PHENOMENON

Epigraphs: Julie Goodman, "Educators Discuss Evolution, with a Wary Eye to Creationism," 6 May 2000, Associated Press Archive at http://nl.newsbank.com. Richard C. Lewontin, Introduction to *Scientists Confront Creationism,* ed. Laurie R. Godfrey (New York: W. W. Norton, 1983), xxv.

1. "Creationism in NZ 'Unlikely,' " *NZ Herald,* 3 July 1986, 14. On the early spread of creationism, on which this paragraph is based, see Ronald L. Numbers, *The Creationists: From Scientific Creationism to Intelligent Design,* exp. ed. (Cambridge, Mass.: Harvard University Press, 2006), 355–68.

2. Numbers, *The Creationists,* passim. Much of the following discussion is extracted from this source, especially chapter 18, "Creationism Goes Global."

3. Ronald L. Numbers, "Creationists and Their Critics in Australia: An Autonomous Culture or 'the USA with Kangaroos'?" *Historical Records of Australian Science* 14 (June 2002): 1–12; reprinted in *The Cultures of Creationism: Anti-Evolutionism in English-Speaking Countries,* ed. Simon Coleman and Leslie Carlin (Aldershot: Ashgate, 2004), 109–23. Regarding the minister of education, see David Wroe, " 'Intelligent Design' an Option: Nelson," *The Age,* 11 August 2005; and Linda Doherty and Deborah Smith, "Science Friction: God's Defenders Target 3000 Schools," *Sydney Morning Herald,* 14 November 2005.

4. Ronald L. Numbers and John Stenhouse, "Antievolutionism in the Antipodes: From Protesting Evolution to Promoting Creationism in New Zealand," *British Journal for the History of Science* 33 (2000): 335–50; reprinted in *Cultures of Creationism,* ed. Coleman and Carlin, 125–44.

5. Debora MacKenzie, "Unnatural Selection," *New Scientist,* 22 April 2000, 35–39, quotation on 38 (more creationists); "God Is Alive," *Maclean's,* 12 April 1993, 32–37, quotation on 35 (53%); Dennis

Feucht, "Canadian Political Leader Advocates Young-Earth," *Research News and Opportunities in Science and Theology* 1 (July–August 2001): 13, 20 (speech); Tom Spears, "Evolution Nearly Extinct in Classroom," *Ottawa Citizen*, 29 October 2000. See also John Barker, "Creationism in Canada," in *Cultures of Creationism*, ed. Coleman and Carlin, 89–92.

6. Tania Branigan, "Top School's Creationists Preach Value of Biblical Story over Evolution," *The Guardian*, 9 March 2002; "Ken Ham Creates Uproar in England!" *Answers Update* 9 (May 2002): 15; Richard Dawkins, "Young Earth Creationists Teach Bad Science and Worse Religion," *Daily Telegraph*, 18 March 2002; editorial, "Outcry at Creationism in UK Schools," *The Guardian*, 12 May 2003 (debauchery); Michael Gross, "US-Style Creationism Spreads to Europe," *Current Biology* 12 (2002): R265–66. See also Simon Coleman and Leslie Carlin, "The Cultures of Creationism: Shifting Boundaries of Belief, Knowledge and Nationhood," in *Cultures of Creationism*, ed. Coleman and Carlin, 1–28, esp. 15–17.

7. "Core Values of Science under Threat from Fundamentalism, Warns Lord May," 30 November 2005, available at www.royalsoc .ac.uk; "Britons Unconvinced on Evolution," BBC News, 26 January 2006, available at http://newsvote.bbc.co.uk.

8. Ulrich Kutschera, "Darwinism and Intelligent Design: The New Anti-Evolutionism Spreads in Europe," 23 (September–December 2003): 17–18 (poll). Martin Enserink, "Is Holland Becoming the Kansas of Europe?" *Science* 308 (2005): 1394. See also Ulrich Kutschera, "Low-Price 'Intelligent Design' Schoolbooks in Germany," *NCSE Reports* 24 (September–October 2004): 11–12.

9. "Creationism in Italy," *Acts & Facts* 21 (March 1992): 5; "Italian Web Site Now Online!" *Answers Update* 9 (April 2002): 19; Silvano Fuso, "Antidarwinism in Italy," www.cicap.org/en_artic/at101152.htm (accessed 19 September 2008), the website of the Italian Committee for the Investigation of Claims on the Paranormal; Massimo Polidoro, "Down with Darwin! How Things Can Suddenly Change for the Worse When You Least Expect It," *Skeptical Inquirer* 28 (July–August 2004): 18–19; Frederica Saylor, "Italian Scientists Rally Behind Evolution," *Science & Theology News* (July–August 2004): 1, 5; "Italians Defend Darwin," *Science* 309 (2005): 2160.

10. "Report from Romania!" *Answers Update* 9 (February 2002): 15; Gabriel Curcubet, Romania Home Schooling Association, to author, 1 June 2005; Misha Savic, "Serbian Schools Put Darwin Back in the Books," *Science & Theology News* (October 2004): 6; Almut Graebsch, "Polish Scientists Fight Creationism," *Nature* 443 (2006):

890–91. See also Maciej Giertych, "Creationism, Evolution: Nothing Has Been Proved," *Nature* 444 (2006): 265. The Serbian minister was subsequently forced to resign.

11. "Second Moscow International Symposium on Creation Science," *Acts & Facts* 23 (July 1994): 2–3; "Creation Science Conference in Moscow," *Acts & Facts* 24 (December 1995): 1–2; "Creationist Curriculum in Russia," *Acts & Facts* 24 (April 1996): 2; John and Svetlana Doughty, "Creationism in Russia," *Impact No. 288,* supplement to *Acts & Facts* 25 (June 1997): i–iv; "Creation International," *Acts & Facts* 31 (2002): 1–2; Massimo Polidoro, "The Case of the Holy Fraudster," *Skeptical Inquirer* 28 (March/April 2004): 22–24; Molleen Matsumura, "Help Counter Creationism in Russia," *NCSE Reports* 19 (No. 3, 1999): 5.

12. "ICR Spanish Ministry Exploding Across the World!" *Acts & Facts* 31 (November 2002): 2; Elaine Brum, "E no princípio era o que mesmo?" *Época,* Edição 346, 3 January 2005 (http://revistaepoca.globo .com/EditoraGlobe/componente), regarding 2004 poll; Jaime Larry Benchimol, "Editor's Note," *História, Ciências, Saúde—Manguinhos* 11 (2004): 237–38; Frederica Saylor, "Science, Religion Clash in Schools around the Globe," *Science & Theology News* 5 (September 2004): 16; Nick Matzke, "Teaching Creationism in Public School Authorized in Rio de Janeiro, Brazil," 14 July 2004 (www.ncseweb.org); Daniel Sottomaior to author, 29 May 2005.

13. Young-Gil Kim, "Creation Science in Korea," *Impact No. 152,* supplement to *Acts & Facts* 15 (February 1986): i–iv; Chon-Ho Hyon, "The Creation Science Movement in Korea," *Impact No. 280,* supplement to *Acts & Facts* 25 (October 1996): i–iv; "Can Creation Science Be Found Outside America?" *Acts & Facts* 30 (March 2001): 4 (largest); "Korea Association of Creation Research Begins West Coast Ministry," *Acts & Facts* 28 (March 1999): 3; Paul Seung-HunYang to author, 24, 28 July and 1 August 2000; Kyung Kim to author, 1 August 2000 (membership, Indonesia).

14. "ICR Book Used in Turkey," *Acts & Facts* 16 (July 1987): 2; Ümit Sayin and Aykut Kence, "Islamic Scientific Creationism: A New Challenge in Turkey," *NCSE Reports* 19 (November/December 1999): 18–20, 25–29; Robert Koenig, "Creationism Takes Root Where Europe, Asia Meet," *Science* 292 (2001): 1286–87.

15. Harun Yahya, *The Evolution Deceit: The Collapse of Darwinism and Its Ideological Background,* trans. Mustapha Ahmad (London: Ta-Ha Publishers, 1999; first published in Turkey in 1997), 1–2; biographical details come from personal interviews with Adnan Oktar and

Mustafa Akyol, 18 December 2000, and from a biographical sketch that appears on www.harunyahya.net: "The Author's Biography: The Story of Adnan Oktar's Life & Ministry." For a fuller account of Harun Yahya and BAV, see Numbers, *The Creationists*, 421–27. On Muslim creationism, see also Seng Piew Loo, "Scientific Understanding, Control of the Environment and Science Education," *Science & Education* 8 (1999): 79–87; and Martin Riexinger, "The Reaction of South Asian Muslims to the Theory of Evolution; or, How Modern Are the Islamists?" Eighteenth European Conference on Modern South Asian Studies, Lund, 5–9 July 2004, at www.sasnet.lu.se/panelabstracts/44 .html.

16. Alexander Nussbaum, "Creationism and Geocentrism among Orthodox Jewish Scientists," *NCSE Reports* 22 (January–April 2002): 38–43, quotations on 39; M. M. Schneersohn, "A Letter on Science and Judaism," in *Challenge: Torah Views on Science and Its Problems,* ed. Aryeh Carmell and Cyril Domb (Jerusalem: Feldheim Publishers, 1976), 142–49. See also Schneerson's essay "The Weakness of the Theories of Creation" (1962) at www.daat.ac.il/daat/english/weakness .htm. On Schneerson's influence on Torah science, see Tsvi Victor Saks's statement at www.torahscience.org/community/saks.html. For Zeiger's views, see the *TSF Newsletter* 3 (December 2003) at www.torahscience .org/newsletter1.html.

17. MacKenzie, "Unnatural Selection," 35–39, plus the cover of the 22 April 2000, issue and an editorial on 3; Interacademy Panel on International Issues, "IAP Statement on the Teaching of Evolution," December 2005, personal copy. See also Gross, "US-Style Creationism Spreads to Europe," R265–66; and Kutschera, "Darwinism and Intelligent Design," 17–18.

MYTH 25: THAT MODERN SCIENCE HAS SECULARIZED WESTERN CULTURE

Epigraphs: Richard Dawkins, *The God Delusion* (London: Bantam Press, 2006), 50. Anthony F. C. Wallace, *Religion: An Anthropological View* (New York: Random House, 1966), 264–65. Cited in Ronald L. Numbers, *Science and Christianity in Pulpit and Pew* (New York: Oxford University Press, 2007), 129.

1. Center for Inquiry, "Declaration in Defense of Science and Secularism," at www.cfidc.org/declaration.html (accessed 9 May 2007).

2. David Martin, "Does the Advance of Science Mean Secularisation?" *Science and Christian Belief* 19 (2007): 3–14.

3. David L. Gosling, *Science and the Indian Tradition* (London: Routledge, 2007), 67.

4. Noah Efron, *Judaism and Science: A Historical Introduction* (Westport, Conn.: Greenwood Press, 2007), 205.

5. Peter L. Berger, ed., *The Desecularization of the World: Resurgent Religion and World Politics* (Grand Rapids, Mich.: Eerdmans, 1999); Ronald L. Numbers, "Epilogue: Science, Secularization, and Privatization," in *Eminent Lives in Twentieth-Century Science and Religion*, ed. Nicolaas A. Rupke (Frankfurt: Peter Lang, 2007), 235–48.

6. John Hedley Brooke, Margaret Osler, and Jitse Van Der Meer, eds., *Science in Theistic Contexts: Cognitive Dimensions, Osiris* 16 (2001).

7. John Brooke and Geoffrey Cantor, *Reconstructing Nature: The Engagement of Science and Religion* (Edinburgh: T & T Clark, 1998), 207–43.

8. Max Caspar, *Kepler* (London: Abelard-Schuman, 1959), 267.

9. Francis Collins, *The Language of God: A Scientist Presents Evidence for Belief* (London: Free Press, 2006).

10. John Hedley Brooke, "Science and Secularization," in *Reinventing Christianity*, ed. Linda Woodhead (Aldershot: Ashgate, 2001), 229–38.

11. Frederick Burkhardt, ed., *The Correspondence of Charles Darwin*, vol. 7 (Cambridge: Cambridge University Press, 1991), 380, 407, 409.

12. John Durant, ed., *Darwinism and Divinity* (Oxford: Blackwell, 1985), 19–20, 28.

13. John Hedley Brooke, " 'Laws Impressed on Matter by the Deity'? The *Origin* and the Question of Religion," in *The Cambridge Companion to the* Origin of Species (Cambridge: Cambridge University Press, 2008).

14. Dawkins, *God Delusion*, 50.

15. T. H. Huxley, "On the Reception of the 'Origin of Species,' " in *The Life and Letters of Charles Darwin*, ed. Francis Darwin, 3 vols. (London: Murray, 1887), 2:179–204.

16. Charles Darwin, *The Autobiography of Charles Darwin, 1809–1882, with Original Omissions Restored*, ed. Nora Barlow (London: Collins, 1958), 87.

17. See James Moore, Myth 16 in this book.

18. Susan Budd, *Varieties of Unbelief: Atheists and Agnostics in English Society, 1850–1960* (London: Heinemann, 1977).

19. Ibid., 107–9.

20. Ibid., 109.

21. Mary Douglas, "The Effects of Modernization on Religious Change," *Daedalus*, issued as *Proceedings of the American Academy of Arts and Sciences* 111 (1982): 1–19.

22. James R. Moore, *The Post-Darwinian Controversies: A Study of the Protestant Struggle to Come to Terms with Darwin in Great Britain and America, 1870–1900* (Cambridge: Cambridge University Press, 1979), 24–29.

23. Bernard Lightman, "Science and Unbelief," in *Science and Religion around the World*, ed. John Hedley Brooke and Ronald L. Numbers (New York: Oxford University Press, 2009).

24. Martin, "Advance of Science," 9.

25. Numbers, *Science and Christianity*, 135.

26. Edward J. Larson and Larry Witham, "Scientists Are Still Keeping the Faith," *Nature* 386 (1997): 435–36.

27. Edward J. Larson and Larry Witham, "Leading Scientists Still Reject God," *Nature* 394 (1998): 313. For a recent survey of 1,646 scientists, see Elaine Howard Ecklund, "Religion and Spirituality among Scientists," *Contexts* 7, no. 1 (2008): 12–15.

28. Martin, "Advance of Science," 5.

CONTRIBUTORS

JOHN HEDLEY BROOKE is Andreas Idreos Professor Emeritus of Science and Religion at Oxford University. His many publications include *Science and Religion: Some Historical Perspectives,* winner of the 1992 Watson Davis Prize of the History of Science Society, and (with Geoffrey Cantor) *Reconstructing Nature: The Engagement of Science and Religion,* originally given as the Gifford Lectures.

LESLEY B. CORMACK is Dean of Arts and Social Sciences at Simon Fraser University. She is author of *Charting an Empire: Geography at the English Universities, 1580–1620,* and coauthor (with Andrew Ede) of *A History of Science in Society: From Philosophy to Utility.*

DENNIS R. DANIELSON is Professor of English at the University of British Columbia. He is the editor of *The Book of the Cosmos: Imagining the Universe from Heraclitus to Hawking* and author of *The First Copernican: Georg Joachim Rheticus and the Rise of the Copernican Revolution.*

EDWARD B. DAVIS is Professor of the History of Science at Messiah College. He is the editor (with Michael Hunter) of a complete edition of *The Works of Robert Boyle* and author of a forthcoming book about the religious beliefs of prominent American scientists in the 1920s.

NOAH J. EFRON chairs the Program in Science, Technology and Society at Bar-Ilan University in Israel and serves as President of the Israeli

Society for History and Philosophy of Science. He is the author of *Real Jews: Secular versus Ultra-Orthodox and the Struggle for Jewish Identity in Israel* and *Judaism and Science: A Historical Introduction*.

MAURICE A. FINOCCHIARO is Distinguished Professor of Philosophy Emeritus at the University of Nevada, Las Vegas. Among his many books are *The Galileo Affair: A Documentary History* and *Retrying Galileo, 1633–1992*.

SYED NOMANUL HAQ is a professor in the School of Humanities and Social Sciences, Lahore University of Management Sciences, Pakistan, and a visiting scholar of Near Eastern Languages and Civilizations at the University of Pennsylvania. He has published extensively in the field of Islam's intellectual history, including the book *Names, Natures and Things: The Alchemist Jabir ibn Hayyan and his Kitab al-Ahjar (Book of Stones)*.

PETER HARRISON is Andreas Idreos Professor of Science and Religion at the University of Oxford, where he is also a fellow of Harris Manchester College and Director of the Ian Ramsey Centre. His several books include *The Bible, Protestantism and the Rise of Natural Science* and *The Fall of Man and the Foundations of Science*.

EDWARD J. LARSON is University Professor of History at Pepperdine University, where he holds the Darling Chair in Law. He received the 1998 Pulitzer Prize in history for his book *Summer for the Gods: The Scopes Trial and America's Continuing Debate Over Science and Religion*; the latest of his many other books is *A Magnificent Catastrophe: The Tumultuous Election of 1800, America's First Presidential Campaign*.

DAVID C. LINDBERG is Hilldale Professor Emeritus of the History of Science at the University of Wisconsin–Madison. A past president of the History of Science Society and a recipient of its Sarton Medal, he has written or edited more than a dozen books, including the widely translated *The Beginnings of Western Science*, which won the 1994 Watson Davis Prize of the History of Science Society.

DAVID N. LIVINGSTONE is Professor of Geography and Intellectual History at the Queen's University of Belfast. A fellow of the British

Academy, he is the author of a number of books, including *Putting Science in its Place* and *Adam's Ancestors: Race, Religion and the Politics of Human Origins*.

JAMES MOORE is Professor of the History of Science at the Open University in England. His books include *The Post-Darwinian Controversies: A Study of the Protestant Struggle to Come to Terms with Darwin in Great Britain and America, 1870–1900* and (with Adrian Desmond) the bestselling biography *Darwin*, now in eight languages.

RONALD L. NUMBERS is Hilldale Professor of the History of Science and Medicine and of Religious Studies at University of Wisconsin–Madison and president of the International Union of the History and Philosophy of Science. He has written or edited more than two dozen books, including *The Creationists: From Scientific Creationism to Intelligent Design* and *Science and Christianity in Pulpit and Pew*.

MARGARET J. OSLER is Professor of History and Adjunct Professor of Philosophy at the University of Calgary. Her books include *Divine Will and the Mechanical Philosophy: Gassendi and Descartes on Contingency and Necessity in the Created World* and the forthcoming *Reconfiguring the World: Nature, God, and Human Understanding in Early Modern Europe*.

KATHARINE PARK is Zemurray Stone Radcliffe Professor of the History of Science at Harvard University. Her many publications include *Secrets of Women: Gender, Generation, and the Origins of Human Dissection* and (with Lorraine Daston) *Wonders and the Order of Nature, 1150–1750*, which won the 1999 Pfizer Prize of the History of Science Society.

LAWRENCE M. PRINCIPE is the Drew Professor of the Humanities at Johns Hopkins University, where he holds appointments in the Department of the History of Science and Technology and the Department of Chemistry. He is the author of *The Aspiring Adept: Robert Boyle and His Alchemical Quest* and (with William R. Newman) of *Alchemy Tried in the Fire: Starkey, Boyle, and the Fate of Helmontian Chymistry*, winner of the 2005 Pfizer Prize awarded by the History of Science Society.

ROBERT J. RICHARDS is the Morris Fishbein Professor of the History of Science and Medicine at the University of Chicago, where he is also professor in the departments of history, philosophy, and psychology. His books include *Darwin and the Emergence of Evolutionary Theories of Mind and Behavior*, which won the History of Science Society's 1988 Pfizer Prize; *The Romantic Conception of Life: Science and Philosophy in the Age of Goethe*; and *The Tragic Sense of Life: Ernst Haeckel and the Struggle over Evolutionary Thought.*

JON H. ROBERTS is the Tomorrow Foundation Professor of American Intellectual History at Boston University. He is the author of *Darwinism and the Divine in America: Protestant Intellectuals and Organic Evolution, 1859-1900*, which won the Frank S. and Elizabeth D. Brewer Prize from the American Society of Church History, and (with James Turner) *The Sacred and the Secular University.*

NICOLAAS A. RUPKE is Professor of the History of Science and Director of the Institute for the History of Science at Göttingen University. Among his books are *The Great Chain of History: William Buckland and the English School of Geology, 1814–1849*; *Richard Owen: Victorian Naturalist*; and *Alexander von Humboldt: A Metabiography.*

MICHAEL RUSE is the Lucyle T. Werkmeister Professor of Philosophy at Florida State University, where he directs the Program in the History and Philosophy of Science. He has written or edited some three dozen books, including *Darwin and Design: Does Evolution Have a Purpose?*, *The Evolution-Creation Struggle*, and *Darwinism and Its Discontents.*

RENNIE B. SCHOEPFLIN is Professor of History and Acting Associate Dean of the College of Natural and Social Sciences at California State University, Los Angeles. He is the author of *Christian Science on Trial: Religious Healing in America* and of a forthcoming monograph that examines changing American understandings of earthquakes and other natural disasters.

JOLE SHACKELFORD teaches in the Program for the History of Medicine and Biological Sciences at the University of Minnesota. An expert on the social and intellectual responses to the chemical, medical, and religious ideas of the Paracelsians, he is the author of *A Philosophical Path for Paracelsian Medicine: The Ideas, Intellectual Context, and Influence of Petrus Severinus (1540/2–1602)*, which received the 2007 Urdang Award from the American Institute for the History of Pharmacy.

MICHAEL H. SHANK teaches the history of science before Newton at University of Wisconsin–Madison. He is the author of *"Unless You Believe, You Shall Not Understand": Logic, University, and Society in Late Medieval Vienna* and of a forthcoming study of the fifteenth-century German astronomer Regiomontanus.

MATTHEW STANLEY is an associate professor at New York University's Gallatin School of Individualized Study. Having recently published *Practical Mystic: Religion, Science, and A. S. Eddington*, which examines how scientists reconcile their religious beliefs and professional lives, he has turned his attention to how science changed from its historical theistic foundations to its modern naturalistic ones.

DANIEL PATRICK THURS teaches science studies in the John William Draper Interdisciplinary Master's Program in Humanities and Social Thought, New York University. He is the author of the recently published *Science Talk: Changing Notions of Science in American Popular Culture*.

INDEX